SOLID-STATE
HIGH-FREQUENCY
POWER

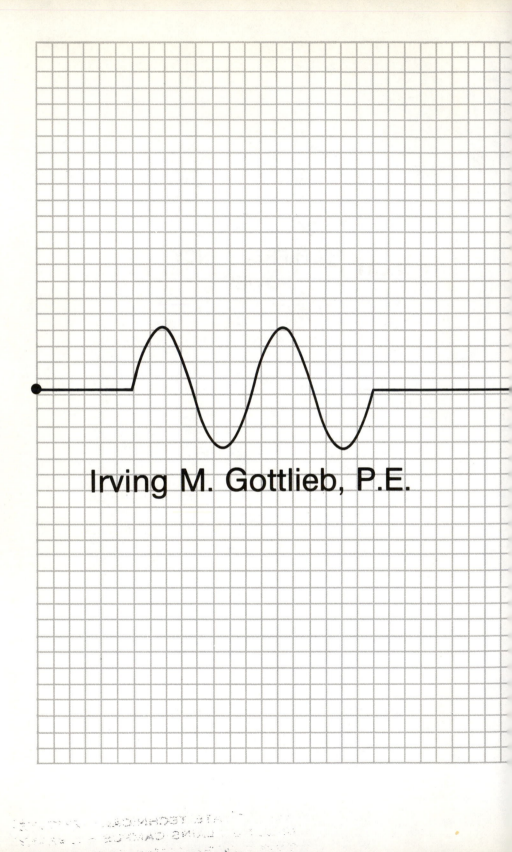

Irving M. Gottlieb, P.E.

SOLID-STATE HIGH-FREQUENCY POWER

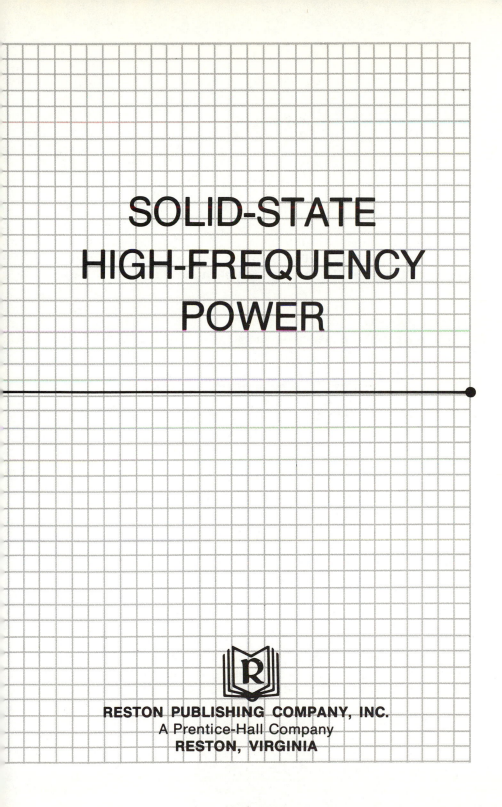

RESTON PUBLISHING COMPANY, INC.
A Prentice-Hall Company
RESTON, VIRGINIA

To my son,
Marc E. Gottlieb

Library of Congress Cataloging in Publication Data

Gottlieb, Irving M.
   Solid-state high-frequency power.

   Includes index.
   1. Power transistors.   2. Amplifiers, Radio frequency.
3. Transistor amplifiers.   I. Title.
TK7871.9.G65   621.3815'28   81-12128
ISBN 0-8359-7048-5   AACR2

Editorial/production supervision and interior design
by Barbara J. Gardetto

© 1982 by
Reston Publishing Company, Inc.
A Prentice-Hall Company
Reston, Virginia 22090

10 9 8 7 6 5 4 3 2 1

Printed in the United States of America

Contents

2
● The Bipolar Transistor in RF Power Applications 42

3
● The Field-Effect Transistor in RF Power Applications 75

4

● Impedance-Matching Networks 103

5

● Applications of Transmission-Line Elements to RF Power Circuitry 153

6

● Low-Power Applications 186

7

● Medium- and High-Power Applications 212

Preface

The use of solid-state devices in the production or processing of radio-frequency energy is one of the noteworthy accomplishments of modern electronics. The dramatic improvements imparted to transmitters, transceivers, space communications, and to hobbyist endeavors share parity with the somewhat more visible achievements involving computers, calculators, stereo, and TV games. Radio frequencies at appreciable power levels also assume importance in such diverse areas as radar, microwave cooking, induction heating, navigation, welding, and numerous scientific and industrial processes. The need for a treatise dealing with device and circuit techniques in this now-evolving field of electronic technology is, in the author's opinion, overdue.

A peculiar aspect of solid-state radio-frequency power is that it has *not* displaced the vacuum tube. Indeed, much evidence suggests that it is not likely to, discounting, of course, the impact of unforeseen developments. This situation merely whets the appetites of designers and experimenters, for it is clearly the temper of the times to maximally exploit the desirable features of semiconductor electronics. Illustrative of this trend is the present attempt to substitute a solid-state device for the magnetron in microwave ranges. Although this newly popular consumer product is destined to benefit from such change, formidable technical and economic barriers bar the path to progress at the time of this writing. It is well to keep in mind, however, that we have already lived through almost unthinkable advances in the field of solid-state RF power. Thus, when power transistors first became available, the notion of extending their capabilities to handle tens and hundreds of watts in the high-frequency RF spectrum remained an idle dream. And the thyristors now used in induction heating applications and elsewhere, where low-frequency RF at high-power levels is needed, were not easily envisioned, even at a time when it was popularly conceded that this device had attained "maturity."

The technical approach and editorial format of *Solid-State RF Power* are intended to enhance the expertise of engineers and designers, as well as to serve the needs of technicians, radio amateurs, servicemen, hobbyists, and experimenters. The author respectfully points out that such apparently extensive coverage is both practical and natural, because today's solid-state practitioner wears more than one hat. For example, a "radio amateur" may actually function in several, if not all, of the aforementioned occupational slots.

The author gratefully acknowledges the assistance obtained from the following companies: Amperex; Communications Transistor Corporation; Continental Electronics; General Electric; International Rectifier Corporation; KLM Electronics; Motorola; RCA; Siliconix; and TRW Semiconductor.

Irving M. Gottlieb, P.E.
Menlo Park, California
August 1979

1

Transistors Versus Tubes

Inasmuch as vacuum tubes have been, and remain, viable devices used in the processing of RF power, generalized comparisons are made between tube and solid-state technologies. Hams, engineers, and others involved in RF work are accustomed to tube circuits and tend to evaluate solid-state applications in terms of their tube experience. And even though solid-state RF applications tend to displace many tube implementations, there is also a well-defined trend to use tubes and transistors cooperatively.

● What Is Radio-Frequency Power?

From a purist's point of view, all alternating currents regardless of frequency radiate *some* energy into space. So, strictly speaking, radio-frequency (RF) power need not be constrained by any limits imposed on frequency. Practically, however, radiation is negligible at the 60-hertz (Hz) power-line frequency. It has been found that by the time we get to 10 kilohertz (kHz), it is possible to detect the radiant energy across the ocean if a sufficiently large transmitting antenna is used. Actually, even lower frequencies would suffice for such long-distance signaling. However, 10 kHz

1

is an audible frequency, and lower frequencies could, among other things, introduce interferences with the audio-frequency information we might wish to convey on such radiant energy. And, of course, the physical size of efficient antennas discourages the use of such very low frequencies. All things considered, we can say that the RF portion of the electromagnetic spectrum starts at 10 kHz. Although such a low frequency may no longer find much use for communicatións purposes, 10-kHz RF power generators are useful in industry for induction heating, case hardening, and in similar processes. The designers and operators of such equipment know they are involved with a unique domain of electrical power because of the shielding precautions required to prevent radiation. Such inadvertent radiation can interfere with communications or other sensitive equipment.

As we go higher in frequency, we find that the radiation of energy into space is more readily accomplished with practically sized antennas; indeed, at tens and hundreds of megahertz (MHz), appreciable radiation may take place from the antenna effects of resonant circuits, and even from connecting wires.

In the microwave region of the spectrum, a frequency of 100,000 MHz may arbitrarily be defined as the high-frequency limit of RF power, because at higher frequencies the radiant energy begins to manifest itself more as infrared "heat" than as radio waves. All told, then, RF power may involve alternating currents from 10 kHz to 100,000 MHz. This corresponds to the wavelength range of from 30,000 meters (m) to 0.3 centimeter (cm).

In addition to communications applications, RF power is used in radar, for cooking, for induction and dielectric heating in industry, for circuit isolation techniques, and for ionizing gaseous devices. A proposed use is as an intermediate energy medium in the conversion of solar energy from space to terrestrial electrical energy (see Figure 1-1). Unlike 60-Hz power, RF power is no longer produced by rotating machinery. Rather, it is generated by electron tubes and, to an ever-increasing extent, by solid-state devices. The relatively recent advent of such solid-state RF power makes a book such as this timely.

● Why Should RF Power Merit Special Attention?

When the generation and distribution of electricity first attained commercial proportions, the relative merits of alternating and direct current were a controversial subject. It was soon appreciated, however, that each mode had its own domain of applications and that the behavior of alternating current could be just as precisely predicted as that of direct current. And when devices became available to amplify and process voice and sound frequencies, the wonder of it all was taken in stride because these audio frequencies were, after all, the well-understood alternating current,

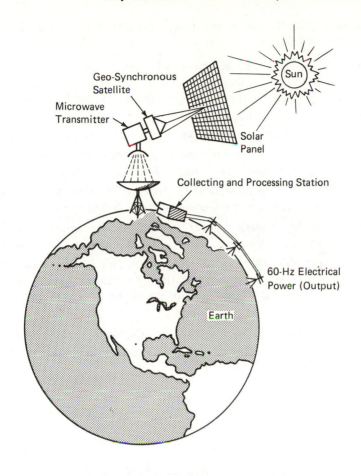

_____ **FIGURE 1-1** _____
Possible RF Power Application of the Future.
Today's technology would require tubes in the transmitter and
solid-state elements at the earth station.

although encompassing higher frequencies than the 50 or 60 Hz used in power engineering. At about the same time, another mode of electrical phenomena began to be commercially exploited—radio-frequency power. Although it could be rightfully said that this merely represented more involvement with alternating currents, for practical purposes it is clear that a whole new branch of electrical technology came into being. Indeed, for many years those who specialized in "power" or in "audio" felt somewhat alienated from the practitioners of the _radio_ art. And the viewpoint was reciprocated!

We may naturally ponder the unique aspects of RF power. These can be tabulated as follows:

- The radiating characteristic of high frequencies is certainly unique. The fact that energy is no longer confined to the wires and components of an ''electric circuit'' is hardly a trivial matter.
- The profound effects of wire length and physical dimensions make necessary a unique expertise for practical implementations.
- The tendency for current flow to concentrate on surfaces requires a modified ''Ohm's law'' approach to dissipative losses.
- The behavior of substances in strong RF fields is dramatically different from that experienced at lower frequencies.
- The ability to guide, reflect, refract, and concentrate RF energy in beams constitutes an art with no parallel at power frequencies.
- The popular design aid of ''breadboarding'' of RF power equipment contrasts with the more rigorous approach of the 60-Hz designer, who generally does not enjoy the luxury of experimentation.

So much for the distinctive *nature* of radio frequency power. Where is it used? Table 1-1 is a partial list of applications of this mode of electrical energy.

TABLE 1-1
Domain of Solid-State RF Power

Aircraft radio	Linear amplifiers
Altimeters	Loran
Amateur radio	Marine radio
Broadband amplifiers	Microwave cooking
Broadcast radio	Microwave communications
Citizen's-band (CB) radio	Mobile radio
Collision avoidance systems	Navigation
Dielectric heating	Plasma generators
Distance measuring equipment	Radar
Frequency-modulated (FM) radio	Radio-sonde service
Garage door openers	Satellite communications
Guidance systems	Television transmitters
Induction cooking	Transponders
Induction heating	Welders
Landing systems	

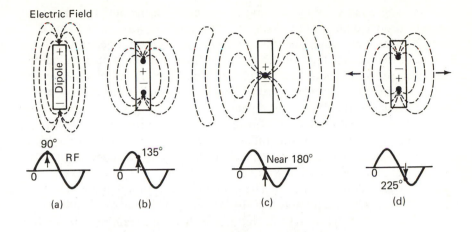

Electric Field

90° RF

135°

Near 180°

225°

(a)   (b)   (c)   (d)

_____ **FIGURE 1-2** _____
Production of Radiant Energy.
Electric lines of force which experience insufficient time to
collapse before reversal of the current cycle close upon
themselves and become detached from the antenna.

● Mechanism of Radiation

Because the salient feature of RF power is *radiation,* we should not
plunge into the actual circuits for producing such power without some ap-
preciation of the nature of this interesting phenomenon. Although this is a
fertile field for exercises in higher mathematics, complex physics, and esoteric philosophic considerations, a simple qualitative explanation suffices for most purposes in practical electronics. In Figure 1-2(a) we see the
electric field around a dipole antenna. We need not concern ourselves
with the length of the antenna, but only that it is driven by an RF generator or transmitter. The changing pattern of the electric field during the cy-
clic excursion of the RF current is due to the fact that there is mutual re-
pulsion between lines of electric force. Thus, they become "pushed"
away from the dipole antenna [Figure 1-2(b)]. This, in itself, does not lead
to detachment from the antenna, for these lines could, conceivably, collapse in reverse fashion from the build-up process. Indeed, this does
occur immediately around the antenna in the region ascribed to the *induction field.* However, if the frequency is high enough, those lines which
have been pushed sufficiently away from the antenna are not given
enough time to collapse before the charges at the ends of the antenna reverse polarity.

Such electric force lines then do the next best thing; they establish
closed paths by closing upon themselves, as shown in Figure 1-2(c). But

once having done so, they become detached from the antenna and continue to propagate into space as an electromagnetic wave at the speed of light. The magnetic component is not shown in Figure 1-2 because the behavior of the electric lines of force suffice for explanation of the vital detachment process. The magnetic lines of force are always at right angles to the electric force lines and therefore can be visualized as always being perpendicular to the plane of the page.

A three-dimensional view of a propagating electromagnetic wave is shown in Figure 1-3. This illustration depicts the magnitudes of the electric and magnetic fields once the radiant energy has detached itself from the antenna via the above-described mechanism. As shown, propagation is to the right, along the X axis. The speed of propagation in the near vacuum of space is approximately 3×10^8 meters per second or 186,000 miles per second. It is slightly less in the atmosphere and can be appreciably less in various dielectric substances such as water or rock. Although refractive effects of the ionosphere may impart even higher speeds to radio waves, present theory restricts any information or energy carried by radio waves to the above-cited speed, that is, the speed of light in free space.

An important aspect of the propagation scenerio is that the electric and magnetic fields wax and wane together even though the force lines representing them are always mutually perpendicular. If we look more closely into this *interdependence* of electric and magnetic fields, we can conclude that one type of force field begets the other, and vice versa. More specifically, a time-varying electric field produces a time-varying magnetic field. In turn, a time-varying magnetic field produces a time-varying electric field. And because *time* is required for such fields to go

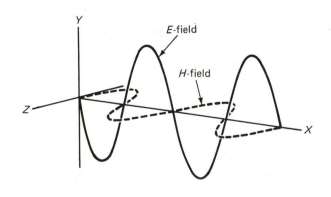

FIGURE 1-3
Spatial Representation of an Electromagnetic Wave.
The mutually perpendicular electric and magnetic fields are
depicted as propagating from left to right along the X axis.

through their respective cycles, a positional advancement of the interaction continually takes place; this constitutes propagation through space.

● Why Reconsider Tubes?

Inasmuch as this text endeavors to impart useful information about RF applications of solid-state devices as power oscillators and power amplifiers, the relevancy of a prelude dealing with vacuum tubes may well be open to question. Seemingly, the momentum of solid-state technology has long pointed in the direction of tube displacement. Why, indeed, should it be profitable to rehash tube circuits and techniques?

It so happens that the circuit principles embodied in the new area of solid-state RF applications derive directly from long-established tube experience. Although it is conceded that it is feasible to "start from scratch" with the solid-state applications, the author feels that a deeper insight into the subject must necessarily stem from acquaintance with or review of tube oscillators and amplifiers. Admittedly, a measure of nostalgia probably permeates this opinion. But for those who are eager to buy the old and become involved with the new, it also happens that tube RF applications are not quite ready yet for burial. In this field, if in none other, the tube retains much vitality; evolving circuit technology often appears in tube designs first, and then is adapted to solid state as devices meeting the contradictory requisites of power capability, frequency, cost, and ruggedness ultimately appear on the market. Tube to solid state represents a continuum in RF power applications rather than an abrupt transition. Evidence of this is seen in mutual-device systems where, for example, solid-state devices are used in the driver stages and tubes are used in the output stage. Actually, *both* semiconductor and tube technology continue to progress. It is clear that the two approaches to RF power can be *either* competitive or cooperative.

● Advantages of Tubes for RF Work

Despite the manifold features of solid-state devices, tubes appear to possess a few advantages. For one thing, tubes can attain much higher power levels in most RF applications. This remains true even though the power capabilities of today's solid-state devices are at levels one hardly dared fantasize just a few years ago. And tubes are much more "forgiving" than are solid-state devices; they do not readily succumb to catastrophic destruction following abuse. Often the tube comes out favorably in a cost-per-watt comparison. (This, however, is no longer construed as an *inherent* feature of tubes.) Until recently, one could say that tube RF amplifiers were easier to drive, but with the advent of the power field-effect transistor, this is no longer necessarily true. Finally, the ability of tubes to operate at high voltages and relatively low currents is in *some* respects advantageous.

● Advantages of Solid-State RF Power Devices

It may appear that involvement with this question is tantamount to revival of a topic long ago resolved. It, of course, is obvious that solid-state devices have size and weight advantages over their tube counterparts. And, it need not be emphasized that dispensing with filament power constitutes an advantage. Perhaps not immediately obvious is the fact that the small size of solid-state devices is advantageous beyond the mere conservation of packaging space. In RF circuitry, particularly at higher frequencies, compact dimensions lead to better circuit performance. Indeed, the physical dimensions of the device, be it tube or solid state, constitute one of the main limiting factors for very high frequency operation. Heat-removal techniques for solid-state devices are more straightforward and are accomplished with simpler hardware. This is because more intimate thermal contact is possible with the heat-developing elements of solid-state devices than with tubes. A big plus feature of solid state devices is that life-span can be characterized by extreme longevity. Moreover, they tend to die at full performances; usually, demise is not preceded by half-good performances. (At the same time, it is well to be mindful of the fact that sudden death is often the price of abuse!)

Because of physical compactness, solid-state RF devices paved the way for the development of transceivers and truly portable transmitters. This development is also facilitated by the voltage compatibility of many solid-state RF power devices with signal-level solid-state circuitry. The physical ruggedness of solid-state elements also enters the picture here and is particularly advantageous in airborne, space, and military applications. And the fortuitous coincidence that solid-state RF power devices can often operate optimally at the dc voltage levels provided by the batteries in automobiles and boats is a blessing which should not go unmentioned. The relative freedom from maintenance is one of the most compelling attributes of solid-state RF power, and this translates into economic advantage when long-term operation is taken into account.

Graphs such as that in Figure 1-4 are often cited to indicate the general power and frequency domains of tubes and solid-state devices. One should refrain from making a too literal interpretation of such curves, for the performance frontiers of all devices are in a state of flux. At any given time, there are always superior performers which are not practically available because they have not been mass produced or are reserved for high-paying customers, such as space programs and the military. Pending an

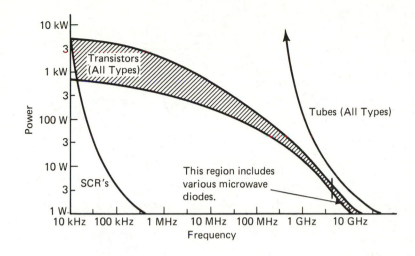

FIGURE 1-4

Approximate Power–Frequency Capabilities of RF Power Devices.

unexpected technological breakthrough, these curves do, however, convey a general picture of the performances pertaining to the indicated devices at a given time. (Also, the plotted power levels are continuous, not pulsed.)

● Class C Tube RF Amplifier

If one understands the basic operation of the class C tube amplifier, most solid-state RF circuits will involve no mysteries. Analogies in performance will, indeed, be closer than one might dare expect. For example, the concept of *dualism* is often expounded to facilitate comparison between the tube and transistor. According to this idea, the two devices are similar, but the tube is best viewed as a voltage-acutuated device, and the bipolar transistor is most easily explained by considering it to be current actuated. However, when we are dealing with class C RF amplifiers, the comparison is considerably simplified, for it turns out that *both* devices behave as if they were actuated by the input current. Yet another similarity has to do with the inherent operation of the class C amplifier. Specifically, it is *not* an amplifier in the sense that it delivers a power-boosted, but faithful, replica of its input signal. The circuit actually performs as a *switch* and is used to shock-excite oscillation in an *LC tank*. Because of this, it really makes no great difference whether the switch is a tube, a bipolar transistor, or a field-effect transistor. Indeed, even an SCR

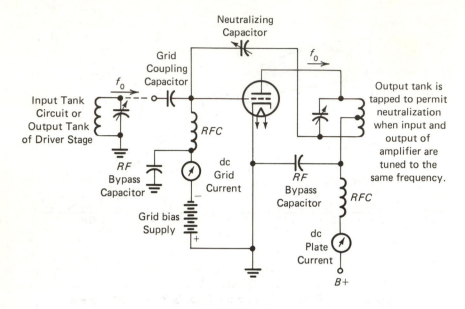

———————————— FIGURE 1-5 ————————————
Representative Tube Class C RF Power Amplifier.
Reconsideration of the operating principles of this basic circuit
provides insight into the newer solid-state systems.

can operate in a similar manner to develop low-frequency RF power in a
resonant circuit.

It should now be evident why some time devoted to tube RF circuits
is more likely to prove rewarding than wasteful. Yet another reason for
such an editorial format is the practical fact that solid-state RF circuitry is
often deployed to replace tube designs. It surely must enhance our expertise to know the nature of the displaced circuit and device.

Figure 1-5 depicts a representative class C RF amplifier using a triode
tube. Analogous output stages are employed in numerous transmitters.
Providing the required power output level is not too great, an ever-increasing number of such transmitters are destined to be manufactured
with solid-state output stages. In any event, a salient feature of the circuit
of Figure 1-5 is the application of sufficient negative grid bias to more than
cut off plate current flow. This being the case, a large enough input signal
must be impressed on the grid to *overcome* the inhibiting effect of the high
negative bias. In actual operation, it is only the positive tips of the input
signal which turn the tube on. Right away we see that plate current flows
only in short duration bursts. Thus, the tube is caused to simulate a switch
which, although synchronized to the input signal, is allowed to be on for

only a small fraction of a complete cycle. The term "amplifier" is clearly a misnomer.

● Basic Operating Features of Class C Amplifiers

When hams and RF circuit engineers made the transition from tubes to solid-state devices, they surprisingly discovered they were right at home with the basic operating features of class C amplifiers. For example, very high efficiency still prevailed, and the counterpart of the plate meter still obligingly dipped to indicate resonance during tune-up. These two operating features were not always well understood during the long domain of the tube. In retrospect, it is easy to perceive that neither feature stemmed directly from the nature of the tube itself, but rather from the *overall* class C circuitry.

The wave-form diagram of Figure 1-6 provides useful insights into class C operation. We note that plate current has a low duty cycle; that is, plate current is off for longer intervals than it is on. Although this suggests the possibility of high efficiency operation, a little study of wave forms reveals an even more important operating characteristic, one which *must* lead to low power dissipation and therefore high efficiency. We observe that, with the output *LC* circuit resonant at the input frequency, plate voltage is minimum during the very time that plate current is maximum. Inasmuch as the product of these two quantities is plate dissipation power, this all-important tube loss is inherently low. Once this basic operating mode of class C tube amplifiers is grasped, it can readily be appreciated that the circuit should provide essentially similar operation regardless of the physical nature of the *switch*. In other words, it should not make a great difference whether the dc pulses which shock-excite the *LC* tank are delivered by a vacuum tube, a bipolar transistor, a field-effect transistor, or another device with a control electrode.

The plate-current dip at resonance is one of the widely known behavior characteristics of class C amplifiers. Yet there is considerable misunderstanding with regard to its cause. Here, again, the wave forms in Figure 1-6 will prove informative. It is evident that the dc plate-current meter must indicate some kind of average value inasmuch as plate current is actually in the form of narrow pulses. From the Fourier theorem of wave composition, such a pulse train consists of a dc component (zero frequency) plus numerous harmonics. The first few harmonics have considerable effect upon the shape of the pulses and, therefore, their average dc value. In other words, the dc component, itself, is the result of the magnitude and phase relationships of the harmonics. Moreover, the first harmonic, which is the *fundamental frequency*, exerts the greatest effect on the pulse wave form. When the *LC* tank is resonated, maximum impedance to the flow of the fundamental-frequency currents prevails. This

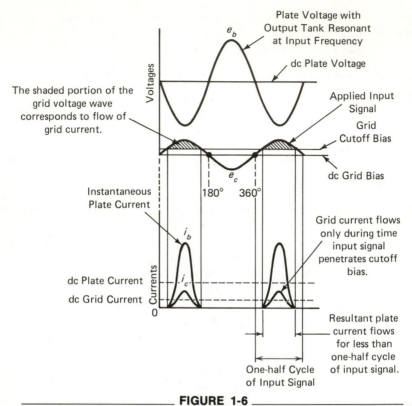

Plate Voltage with
Output Tank Resonant
at Input Frequency

e_b

dc Plate Voltage

The shaded portion of the
grid voltage wave
corresponds to flow of
grid current.

Voltages

Applied Input
Signal

Grid
Cutoff Bias

dc Grid Bias

e_c

Instantaneous
Plate Current

180° 360°

Grid current flows
only during time
input signal
penetrates cutoff
bias.

i_b

dc Plate Current

dc Grid Current

Currents

i_c

0

Resultant plate
current flows
for less than
one-half cycle
of input signal.

One-half Cycle
of Input Signal

FIGURE 1-6

**Voltage and Current Relationships in the Tube Class C Power
Amplifier.**
Obviously, the term "amplifier" is a misnomer, for the plate current
is not a replica of the input signal, but is, rather, a sequence of
short-duration pulses.

alters the shape and magnitude of the plate-current pulses in such a way
that the net dc component, that is, the average value of plate current ac-
tuating the plate-current meter, is greatly reduced. Thus, we have the
somewhat paradoxical situation that attenuation of one of the Fourier
components (the first harmonic, or fundamental frequency) reduces the
net dc component of the plate-current wave form. Accordingly, the plate-
current meter obligingly dips.

● **Linear RF Amplifier**

Not long after tube RF amplifiers began to be deployed for achieving
high power levels, it occurred to an enterprising amateur that his amplitude-modulated transmitter could develop more power by the simple ex-

pedient of inserting a large class C amplifier between the present modulated stage and the antenna. Such a final amplifier did boost the RF power, but it practically destroyed the fidelity of the modulation. With the combination of empirical and analytical techniques, a way was found to produce the sought-after boost in RF power level without distorting the modulating signal. After this was implemented, the circuit of the class C amplifier remained virtually unaltered. What was done to bring about the desired operational mode?

The class C RF amplifier cannot properly handle a modulated signal impressed at its input because the current produced in its output circuit is not proportional to the voltage applied to its input circuit. (This could probably be anticipated from the fact that such an "amplifier" is, in essence, more in the nature of a pulsing or switching system completely dependent upon the energy storage in its output tank circuit for the production of undamped, or constant-amplitude, RF waves.) It turns out that *some* aspects of the class C amplifier can be retained and still have the requisite proportional amplification needed to properly boost the power of modulated waves. Specifically, if the input excitation consists of half-cycles of the RF wave, this important operating condition will occur. That is, instead of supplying say, 90° to 120° of the RF sine wave to the amplifier's input, we impress 180°, or one-half cycle. To accommodate this change, the negative grid bias must be changed; it must be lowered in order to enable the tube to operate as a proportionate, or linear, amplifier over the entire half-cycle of impressed RF. Thus, the circuit itself can conceivably remain unchanged. We are reminded here of class B audio amplifiers, but the comparison is faulted by the strange fact that *one* tube suffices to produce essentially undistorted output of the modulated RF wave. How can this be so, when it is obvious that severe distortion would accrue in a class B audio amplifier if one of its push-pull power tubes were removed from its socket?

The resolution of this paradox lies in the tuned *LC* tank circuit of the class B RF amplifier. Its energy storage, its tendency to "ring," supplies the missing half of the RF cycle. This enables the tube to impart linear amplification of the modulated wave. The class B RF amplifier thus retains the salient feature of the class C amplifier in that a tuned circuit supplies output power when the input drive no longer is present. At the same time, the plate current is zero with no input so that the low operating efficiency of class A operation is circumvented. The theoretical peak efficiency of a class B linear amplifier is 78.5%; that of a class A stage is 50%.

● More-Practical Linear Amplifiers

In practice, it is found desirable to operate the linear RF amplifier in the class AB region, rather than "purely" in class B. This improves the

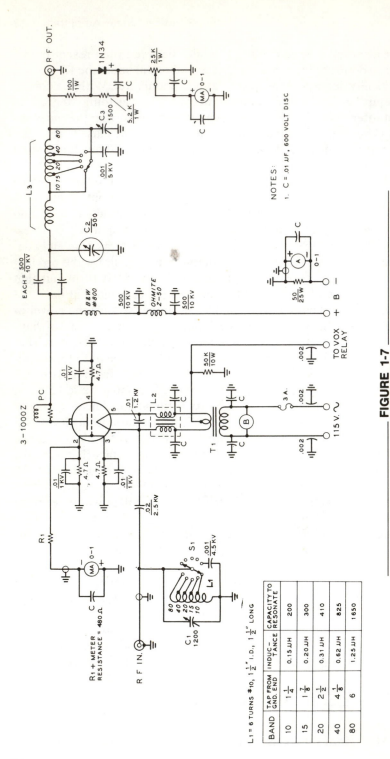

FIGURE 1-7

Typical Circuit of High-Power Grounded-Grid Amplifier.
Numerous radio amateurs, as well as designers of industrial RF equipment, have derived much of their experience from circuits such as this.

BAND	TAP FROM GND. END	INDUC-TANCE	CAPACITY TO RESONATE
10	$1\frac{1}{4}$	0.15 μH	200
15	$1\frac{7}{8}$	0.20 μH	300
20	$2\frac{1}{2}$	0.31 μH	410
40	$4\frac{1}{8}$	0.62 μH	825
80	6	1.25 μH	1650

$L_1 = 6$ TURNS #10, $1\frac{1}{2}$" I.D., $1\frac{1}{2}$" LONG

14

linearity and generally holds true for solid-state, as well as tube, stages. The linearity improvement stems from two sources. First, the dynamic transfer characteristic of the amplifying device is "straightened out." Second, a more constant load is presented to the driver stage. Of course, the "blend" with class A characteristics degrades the operating efficiency somewhat. In actual practice, it proves feasible to achieve a good compromise. This is often brought about by adjusting the bias so that *some* plate current exists without any RF drive. Such an *idling* current is, however, much smaller than would exist if the stage were biased for class A operation. The overall result is linearity approaching that of a class A amplifier and operating efficiency not drastically reduced from that attainable with class B operation.

Even better results are generally forthcoming when the tube is operated in the grounded-grid configuration. Here the control grid is at ground potential (for RF), and the drive is applied to the filament or cathode. At low and medium frequencies, say up to 20 or 30 MHz, the electrostatic shielding action of the grounded grid often stabilizes amplifier operation so that neutralization can be dispensed with. And even when neutralization is used, less critical adjustment and extended broadband operation generally result. Although the impedance presented to the driver stage is low, it tends to be more constant than with grid-driven circuits. The low imput impedance of the grounded-grid stage actually makes it more convenient to use coaxial cable between the driving stage and the final amplifier. A modern high-power grounded-grid amplifier, such as shown in Figure 1-7, can be excited from a solid-state driver.

Although the input to the grounded-grid amplifier is a *power-consuming circuit*, most of the power thus consumed from the driver stage is not dissipated as loss. Rather, it reappears in the output circuit and actually adds to the output power. Typically, 10% of the output power actually derives from *feedthrough* from the driver stage. This is not of consequence in CW, FM, or SSB operation, but will prevent 100% modulation of an AM signal. The grounded-grid circuit is usually a more effective amplifier for the VHF and UHF regions than the "conventional" grounded-cathode configuration. Familiarity with tube grounded-grid amplifiers pays dividends when working with grounded-base transistor circuits; many operating characteristics are quite similar for the two devices in these circuit configurations.

● Transistor Version of the Grounded-Grid Amplifier

Seemingly, a transistorized counterpart of the grounded-grid amplifier would be a "natural," if for no other reason than that neither filament transformer nor filament chokes would be required. One could also be enthusiastic over the better frequency response inherent in common-base circuits as compared to the common-emitter configuration. Also, we may have been accustomed to see superior linearity in power transistors characterized for audio-frequency service when such transistors are operated in common-base amplifiers. (Except for certain servo applications, common-base power amplifiers are now rarely encountered.) Although common-base RF amplifiers (and oscillators) are used, especially at VHF, UHF, and microwaves, most transistor power amplifiers, whether class A, B, C, or "linear," are configured about the common-emitter circuit.

The main reason we do not commonly find transistor versions of the grounded-grid amplifier is that the input impedance of such an arrangement would be so low that practical difficulties would attend the design of an input matching network. Also, the linearity of RF power transistors generally offers little, if any, advantage when operated in the common-base circuit. And, whereas a tube tends to have better RF stability in grounded-grid than quite grounded-cathode circuits, the situation prevailing with transistors may be different; that is, RF stability is often readily attained in common-emitter, but not necessarily in common-base, amplifiers. Note that although the grounded-grid amplifier generally dispenses with neutralization, the common-emitter amplifier generally requires no neutralization. Inspection of many common-emitter amplifiers shows that neutralization is usually *not* required. This is because of low Q "tank" circuits, deliberately introduced emitter degeneration, and the low gain of transistors at high frequencies.

There are, however, similarities between the tube and transistor linear amplifiers. Both are operated as close to class B as is consistent with acceptable linearity. In practice, this usually means that both tube and transistor "linears" operate in the class AB region. That is, both consume small dc idling currents when there is no RF input.

What has been said about transistor versions of the grounded-grid amplifier applies primarily to the bipolar transistor. The relatively new power FET may find service in the analogous circuit (common gate) to the grounded-grid configuration. However, thus far the linearity displayed in common-gate arrangements does not provide much incentive to simulate tube practice.

● Solid-State Approach to High Power

A variety of RF power transistors are available which can develop power outputs of 50, 75, and 100 watts (W) and even more. Many of these transistors can be readily accommodated in broadband circuit arrangements so that the 1.8- to 30-MHz frequency range can be handled without tuning adjustments. The power supply required generally ranges from 12 to 28 volts (V). This is respectable power for many purposes, and somewhat more power is available via push-pull and parallel implementations.

But, suppose we want power at the 1-kilowatt (kW) level. The tube-oriented amateur or designer would probably consult a semiconductor firm's catalog in quest of an appropriate giant transistor. The odds are against finding one. Although kilowatt transistors have been made for military and industrial RF applications, it is just as well that they are not a common commodity on the market.

The practical problems attending the use of such a giant transistor are far from trivial. Conductors, coils, and RF chokes capable of carrying the tremendous dc current would be so massive as to pose serious construction problems. It is not merely a matter of attaining safe current-carrying capacity, but rather of avoiding power-deleting voltage drops. At current levels of, say 150 amperes (A), "small" resistances on the order of several tens of milliohms would seriously impair the operating efficiency. Aside from this annoying situation, it might not be such a good idea to depend on a single device. Thinking of the ease with which lower-power RF transistors can often be destroyed during experimentation or tune-up, it could be less than pleasant to blow out one of the giant units. It appears wiser to try to achieve high power by somehow combining an appropriate number of smaller transistors. Paralleling is generally not too desirable with bipolar transistors. We quickly get into troubles with current hogging; besides, such a combining method does not represent good RF practice at the higher frequencies. This is just as well, for if it were feasible to parallel a large number of 80- or 100-W transistors to develop a kilowatt output level, we would be back where we found ourselves during our contemplation of the single giant transistor. Push-pull circuits are effective techniques for at least doubling power, and in some, but not all, cases a push-pull and parallel arrangement of four transistors can be successfully worked out. Such a scheme would probably yield about three and one-half times rather than four times the available power from one transistor. There must be a better way!

Fortunately, there is another approach. Power-combining circuits can be implemented in which two, three, or four pairs of push-pull transistors may have their individual power outputs summed in the load. The block diagram of Figure 1-8 depicts such a technique for two separate push-pull amplifiers. Load power is four times the contribution of any of the individual transistors. The power divider (also called a *splitter*) and the power combiner are similar to one another and are special *hybrid* transformers. They have their windings arranged and connected so that the amplifier inputs are isolated from one another in the case of the divider, and the amplifier outputs are isolated from one another in the case of the combiner. These transformers are commercially available.

The power combining concept depicted in Figure 1-8 can be extended to produce even greater power from commercially available RF power transistors. For example, divider and combiner hybrid transformers can

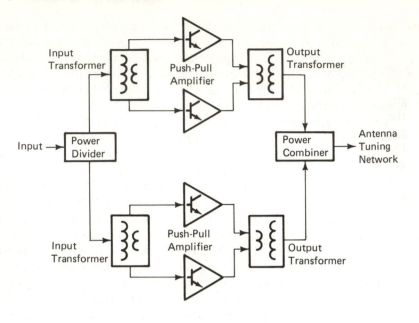

_____ **FIGURE 1-8** _____
Power-Combining Scheme for Two Identical Push-Pull Amplifiers.
Each push-pull amplifier is complete in itself and operates
independently of the other. The antenna load receives the summed
power of these amplifiers or four times the output power of a
single transistor. The power divider and combiner are special
transformers.

be designed to have four ports, as illustrated in Figure 1-9. This enables
the summing of power from *four* separate RF power amplifiers. Interestingly, each RF power amplifier shown in Figure 1-9 can be almost any amplifier which can stand alone and perform well. For example, each of
these amplifiers can represent an entire subsystem, such as the entire arrangement of Figure 1-8. A practical and economical feature of this technique is that the eight push-pull amplifiers that would thereby be deployed
in the technique of Figure 1-9 would be of identical circuitry and construction. Yet another feature is that the loads of either Figure 1-8 or 1-9 continue to be supplied with proportionate power if one or more of the individual amplifiers fail. Such operational redundancy would certainly be
conspicuous by its absence if we were to use a single giant transistor.

● Wilkinson Power-Combining Technique

The Wilkinson power-combining technique provides a simple and straightforward method of combining the power capabilities of two amplifiers where broadbanded operation is not needed. The scheme is shown in Figure 1-10. The elements used are quarter-wave transmission lines; four of these are required. Although initial inspection of the arrangement might suggest push-pull operation, the amplifiers are essentially paralleled. But, unlike ordinary paralleling, there is considerable electrical isolation between the two amplifiers.

Assuming a 50-ohm input source, a 50-ohm load, as well as 50-ohm input and output impedances of the amplifiers, the quarter-wave lines

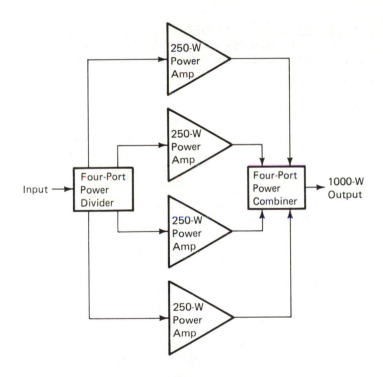

FIGURE 1-9

**Extended Use of Combining Technique for Producing Still
Greater Load Power.**

Here the power-dividing and power-combining transformers each
have four ports. Moreover, each 250-W RF power amplifier can,
conceivably, be a complete subsystem, such as shown
in Figure 1-8.

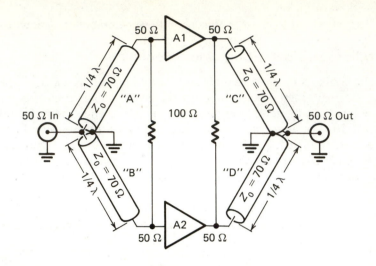

_____FIGURE 1-10_____
Wilkinson Power-Combining Technique.
Under ideal conditions, the 100-ohm resistances dissipate no
power.

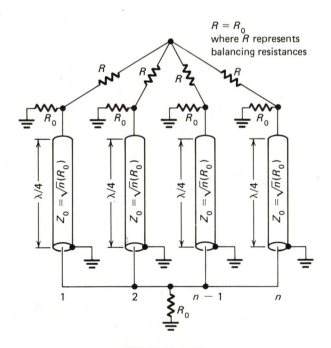

$R = R_0$
where R represents
balancing resistances

_____FIGURE 1-11_____
Wilkinson Divider Combiner Network for Use with n Amplifiers.
R_0 represents the source and load impedances, generally 50 ohms.

should have a characteristic impedance, Z_0, of $50 \times \sqrt{2}$, or approximately 70 ohms. This being the case, one can show that matched impedance conditions prevail. For example, the line A associated with the input of amplifier A1 steps its input impedance of 50 ohms up to 100 ohms at the input terminal of the pair of amplifiers. This comes about from the impedance relationships of the quarter-wave transformer in which $Z_0 = \sqrt{Z_1 \times Z_2}$. Substituting the relevant values, we have $70 \cong \sqrt{100 \times 50}$, Z_1 and Z_2 being the terminating impedances of the line. Inasmuch as two such lines join at the input of the amplifier pair, the actual input impedance seen by the driver or source is 100/2, or 50 ohms. The same basic reasoning applies to the output circuit involving the amplifier outputs, lines C and D, and the load. Note, however, that in the input circuit we have power splitting or dividing, whereas in the output circuit power *combining* takes place.

When the impedance situation is as described, no power dissipation occurs in the 100-ohm resistances. If ideal impedance conditions do not exist, these resistances will absorb the reflected power that will thereby be produced.

● Use of the Wilkinson Scheme with More Than Two Amplifiers

The hybrid networks involving two quarter-wave lines for summing the powers of a pair of amplifiers are actually a specific application of a more generalized technique. Any number of amplifiers can be handled in this manner; it is merely necessary to extend the scheme to accommodate the amplifiers involved. Figure 1-11 shows the more generalized arrangement for use with n, that is, any number of amplifiers. Again, the network can be deployed either as a power divider or power combiner. Note that the configuration reduces to that of Figure 1-10 when $n = 2$. R_0 represents the source (driver) and load impedances and is usually 50 ohms. The characteristic impedance of the lines, Z_0, changes with the number of lines employed. Other parameters and considerations remain the same, regardless of the number of lines. Generally, a 25% bandwidth is attained, and the isolation between ports is on the order of 25 decibels (dB).

The block diagram of Figure 1-12 shows the generalized method of summing the outputs of any number of amplifiers with the use of power dividers and power combiners having the basic configuration depicted in Figure 1-11. Interestingly, the amplifiers themselves can be of any format. That is, they can be configured around single transistors, point-to-point paralleled transistors, push-pull transistors, or power-combining systems such that we have hybrid systems within hybrid systems. Generally, all amplifiers from A1 through An are identical, but it is conceivable that am-

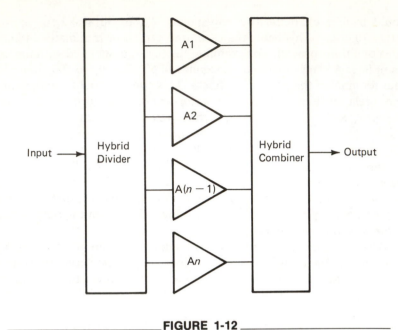

FIGURE 1-12
Wilkinson Hybrid System with *n* Amplifiers.
General power-combining situation is depicted by the block
diagram.

plifiers of diverse formats can have their powers summed in the load via this scheme.

As described, this power-summing method is suitable for narrowband applications. However, considerable broadbanding can be accomplished by using stepped stripline quarter-wave elements, or by cascading techniques in which the lines become odd numbers of quarter-wave lengths beyond 1. Both the application and the mathematical relationships may become unwieldly, however.

● Amplifier Module

Amplifier modules are relatively new, but are already a widely used solid-state RF product. Although various fabrication techniques are used, the common feature of these power amplifiers is that installation consists merely of inserting the unit between the driver (usually the previously used final amplifier) and the antenna, and providing a dc source of power, usually a nominal 12 V. They are sufficiently broadbanded for their intended purpose so that no tuning adjustments need be made. Indeed, there is generally no external adjustment or tuning provision, other than a toggle switch or two for turning on and off the dc power and selecting the operational mode (class B or AB for single sideband, class C for CW or FM). These modules have thus far found greatest application in the VHF and UHF bands where stripline "tank" circuits enable predictable performance from mass-produced units and contribute to stable operation. A large market for these amplifier modules has been owners and operators of transceivers seeking a convenient and inexpensive way to boost power. Power outputs from 15 to 150 W have been commonly available.

Fabrication techniques employ various blends of hybrid and discrete methods, with the heat sink often being the most evident construction feature (see Figure 1-13). Manufacturers vie with one another to provide various performance features. Among these are output harmonic filters, integral preamplifiers for easy drive, transmit-receive relays, jacks for monitoring performance, and, above all, efficiency. These amplifier modules are highly useful for mobile operation and are similarly being welcomed for deployment in marine and aircraft applications.

Some amplifier modules are intended for use in MATV/CATV installations. These generally have power output levels of a fraction of a watt to several watts and may exhibit substantially flat response from 40 to 300 MHz. With somewhat lower power ratings, there are other amplifier modules which are broadbanded from 1 to 250 MHz or from 10 to 400 MHz. It is remarkable that, despite the sophisticated technology and

FIGURE 1-13
Typical RF Power Amplifier Module.
Not visible are input and output BNC connectors for 50-ohm coaxial cable.
(Courtesy of KLM Electronics)

careful design incorporated in RF amplifier modules, they can almost be treated as just another component insofar as concerns installation and operation.

● Low-Power RF Amplifier Module

Hybrid-constructed amplifier modules are useful for driver, pre-driver, and signal-processing functions in more extensive systems. Their deployment saves engineering development time, conserves space, and often provides optimized performance features not readily attained with discrete circuit layouts. Their consideration is worthwhile in light of the fact that most circuit schematics of power-output amplifiers merely depict an IN terminal, thus evading the problem of excitation.

The physical appearance of such a module, the Motorola MHW591, is shown in Figure 1-14. The maximum dimensions are such that the module occupies a volume of less than 1.2 cubic inches. It nonetheless contains eight transistors and three transformers. An alumina substrate is employed with thin-film techniques which include nichrome resistors, gold strip-line elements and interconnections, molybdenum heat-spreaders (for thermal conductivity), and chip capacitors.

The output power is on the order of 1 W and linearity is exceptionally

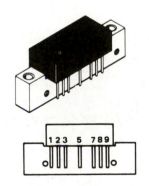

PIN   1. RF INPUT
        5. V$_{DC}$
2,3,7,8. DC AND RF GROUND
        9. RF OUTPUT

_____ **FIGURE 1-14** _____
Example of Hybrid-Constructed RF Amplifier Module.
Intended for driver, and other low-power applications, such
modules take up little space, involve high reliability, and often have
certain optimized performance features.
(Courtesy of Motorola Semiconductor Products, Inc.)

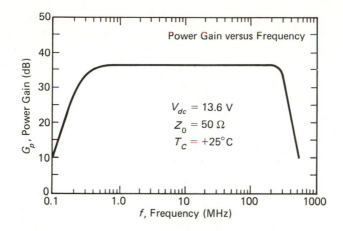

_____ **FIGURE 1-15** _____
**Broadband Frequency Response of the Motorola MHW591 Hybrid
Amplifier Module.**
Such optimized performance features are not readily attained via
the more traditional construction techniques.
(Courtesy of Motorola Semiconductor Products, Inc.)

good. Its overall characteristics prevail when the module is used with 50-100-ohm impedance levels. Probably the salient feature of this particular amplifier module is its inordinately broadbanded response. This is shown in Figure 1-15. Such a module is, itself, a well-engineered subsystem, and one is hard pressed to identify its counterpart in RF designs using tubes. As might be surmised, the uniformity, stability, and reliability of hybrid modules of this type are especially noteworthy.

● Reliability

The reliability of a transmitting-type vacuum tube is, of course, subject to many variables and vagrancies. However, we are reconciled to the fact that its life span is primarily governed by the electron-emission ability of its filament or cathode. In other words, much as with an automobile tire, the device is eventually "used up." And, as with the tire, its functional usefulness generally diminishes from the moment it is pressed into service. Those making the initial transition to solid-state RF power often tend to assume that concern with life span is an archaic notion, and maintenance-free operation is the natural mode of performance with modern silicon devices. Such an assumption may, indeed, appear justified in many practical situations. On the other hand, disappointment may be in store for the subscriber to the premise of indefinitely great longevity for solid-state RF power devices. For besides being vulnerable to catastrophic destruction from transients, "hot-spotting," overdriving, and the like, semiconductor components actually have built-in wear-out mechanisms.

One wear-out mechanism is metal migration. With high current density and high temperatures, the metallic ions in the interconnect or wire bonding system actually are displaced. We are reminded somewhat of electroplating, where ions in a solution are deliberately imparted motion by an electric current. This phenomenon is relatively slow in metals, but eventually produces whiskers, hillocks, and voids. Failure in the device then manifests itself as an internal short or open circuit. Gold is better than aluminum metallization in delaying the onset of this failure mode, and monometallization is generally superior to fabrication processes using more than one metal. (This, however, is influenced by the nature of the metals which may be in contact with one another.)

A closely related failure results from corrosion of metallic bonds, contacts, and interconnects. The semiconductor material, itself, may contribute to such failure.

Another well-documented failure mode in all power semiconductors is thermal fatigue of leads and interconnect metallization. This represents accumulated mechanical strain from the kind of temperature cycling which occurs when the device is first turned on or finally turned off.

Certain abusive conditions during operation or adjustment can shorten the natural life span without necessarily incurring any of the preceding failure modes. Included are overdrive, particularly where the emitter-base diode is driven into avalanche or zener breakdown, high VSWR in the load, voltage transients from the dc supply, parasitic oscillations, and high junction temperatures.

Figure 1-16 shows failure probabilities for a family of RF power transistors as a function of junction temperature. The sample calculation depicts a mean time to failure of about 26 years for typical operating conditions for one of the types. This is a long expected life span for hobbyist and consumer applications, but not necessarily for space or military uses. And one need not be particularly skilled at extrapolation to interpret lifespans of 5 or 10 years for other operating conditions. Considering that the chart predictions are merely weighted averages and that the transistor is always subject to random deviations appreciably removed from the "mean" probability value, it is obvious that the subject of solid-state reliability deserves consideration. This is especially true because solid-state devices usually suffer catastrophic destruction when they fail. From a maintenance point of view, one cannot rely upon any internal time clock, as with thermionic emission in a vacuum tube.

Although only the highlights of a very extensive subject has been touched upon here, there is one overwhelmingly important factor to keep

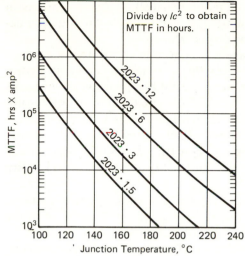

Divide by I_c^2 to obtain MTTF in hours.

The curves pertain to four transistors of the MRAL 2023 family.

Example of MTTF for MRAL 2023.12 Transistor

Where P_o = 12 W

P_{in} = 2.4 W

V_{CC} = 22 V

η_c = 40%

T_{flange} = 70°C

$$I_c \cong I_E = \frac{100\ P_o}{\eta_c \times V_{CC}} = 1.36\ \text{A}$$

$$P_{diss} = P_{in} + V_{CC}I_c - P_o = 20.40\ \text{W}$$

$$T_{junc} = T_{flange} + \theta_j F \times P_{diss} = 161.4°\text{C}$$

$$\text{MTTF} = \frac{4.3 \times 10^5\ \text{hr amp}^2}{I_c^2} = 232.482\ \text{hr}$$

$$= 26.5\ \text{yr}$$

FIGURE 1-16

Typical RF Power Transistor Failure Prediction Chart.
The accompanying sample calculation shows that good, but not
infinite, longevity may be expected.
(Courtesy of TRW Semiconductor)

in mind in the interest of life span. This factor is junction temperature.
Although collector-base junction temperature is implied, it is not to be inferred that high operating temperature is of benign consequence in the
emitter-base junction. However, in most applications, the power dissipation in the emitter-base junction is much less than in the collector-base
junction. Therefore, the temperature of the emitter-base junction tends to
follow that of the collector-base junction. (When hot-spotting, or base-

emitter avalanche, occurs, instantaneous or small-area temperatures in the emitter-base junction may assume destructive levels without appreciably affecting the average junction temperature.)

Summarizing, the insurance agent may well have a different conception of semiconductor longevity than the complacent circuit designer. Transistors are not as forgiving as are tubes; it behooves us to devote as much time to heat removal as to circuitry to spare them from power-supply transients, never to apply excessive drive power, and to try to provide a low VSWR matched load at all times. A final suggestion, one that all too often derives from painful experience, consists of procuring a device in which the manufacturer takes pride in quality control and reliability screening.

● Test Circuits

Most manufacturers of solid-state RF power devices provide test circuits in conjunction with parameter values, tables, and graphs. Such circuits depict typical situations for which the technical information is valid. Both maker and user benefit from these circuits. The maker is largely spared from allegations of unrealistic performance, and the user is provided at least initial guidance predicated upon good engineering practice.

A typical test circuit is shown in Figure 1-17. Here the manufacturer illustrates a simple, yet sophisticated application for a unique product, the JO 2058 transistor, which is specially designed to operate in the common-base configuration and to produce clean 100-W pulses of 450-MHz energy. Anyone with a modicum of experience in high-frequency and pulse techniques and who has worked with RF at high power levels and low impedances can appreciate the likelihood of going astray without such basic "navigation." Indeed, many designers find it is good practice to spend a little extra time ferreting out a manufacturer's test circuit which closely complies with the requirements of the application at hand. Minor modifications will then prove easier and safer to incorporate than will major ones. This is much more true with transistors than with tubes. Most RF power transistors are optimized for best performance under specific circuitry and operating conditions; forcing them to perform under other circumstances can, at best, result in diminished life span and, at worst, produce instant catastrophic destruction.

● Layout Diagrams and Printed Circuit Board Art

At the higher frequencies, a connection diagram is not sufficient information to enable reproduction of the test circuit. This is especially so at VHF, UHF, and microwaves where stripline techniques are used as format for the elements in the impedance-matching networks. It is, therefore, also commonplace for the manufacturer to back up test circuits with layout drawings and with printed circuit (PC) board artwork. Figure 1-18 illustrates such pictorial data for the test circuit of Figure 1-17. The importance of this procedure stems from the fact that the impedance, reactance, and Q of the stripline elements are governed by their physical dimensions and by the nature and thickness of the PC board. Thus, the oft-repeated admonition to keep leads short in RF circuitry is not, in itself, sufficient for reproducibility of performance where the "leads" are deliberately designed to yield circuit functions.

● Amplifier Chains

In solid-state RF power, a practice has come into prominence which although not unknown in vacuum-tube techniques, was not commonplace. This pertains to device manufacturers's concern with the *system*,

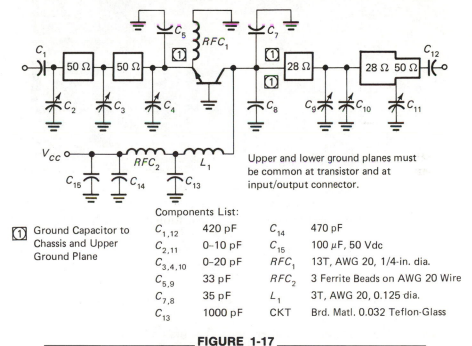

Components List:

$C_{1,12}$	420 pF	C_{14}	470 pF
$C_{2,11}$	0–10 pF	C_{15}	100 μF, 50 Vdc
$C_{3,4,10}$	0–20 pF	RFC_1	13T, AWG 20, 1/4-in. dia.
$C_{5,9}$	33 pF	RFC_2	3 Ferrite Beads on AWG 20 Wire
$C_{7,8}$	35 pF	L_1	3T, AWG 20, 0.125 dia.
C_{13}	1000 pF	CKT	Brd. Matl. 0.032 Teflon-Glass

1 Ground Capacitor to Chassis and Upper Ground Plane

Upper and lower ground planes must be common at transistor and at input/output connector.

_____ **FIGURE 1-17** _____

Typical Manufacturer's Test Circuit.
Such test circuits often provide the basic design information needed to implement a device. This circuit is that of a 100-W pulse amplifier intended for production of 1-ms pulses of 450-MHz energy. The transistor is the JO 2058.
(Courtesy of TRW Semiconductor)

Test Circuit — Layout

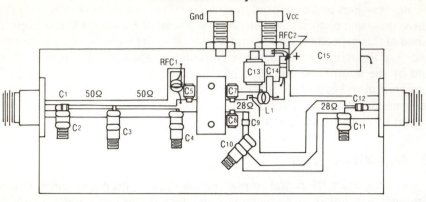

Test Circuit — Artwork

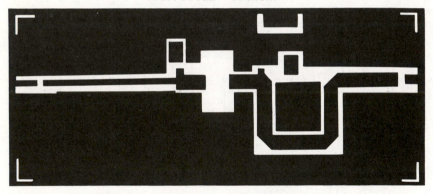

_____**FIGURE 1-18**_____
Board Layout and Artwork for the Test Circuit of Figure 1-17.
The physical and geometrical aspects of such a circuit are as
important as the schematic diagram.
(Courtesy of TRW Semiconductor)

as well as the individual amplifier stage. To this end, "amplifier chains,"
or "suggested lineups," are often included in the specifications literature.
Figure 1-19 is a typical example of such information. From the user's
viewpoint, this can save considerable intellectual and empirical effort.
From the vantage point of the maker of semiconductor devices, any procedure undertaken to enhance practical implementations of his products
is likely to prove rewarding in the marketplace.

● Low-Frequency Applications

We may arbitrarily say that the 10- to 500-kHz region comprises the low-frequency portion of the RF spectrum. Rationale for such classification is simply that the preponderance of technical literature addresses itself to techniques pertinent to much higher frequencies—from several megahertz through the far-microwave region. Low frequencies are readily generated and processed by traditional oscillator-amplifier lineups via the use of power transistors. Below 30 kHz, it often is the case that SCR's are more desirable than transistors, especially for high-power final amplifiers. Many designs make use of class D, rather than class C or linear operation. Here the output voltage wave is square and the duty cycle is ideally 50%. This leads to very high efficiency, as well as better reliability. Such operation is feasible because of the excellent performance of harmonic filters in this frequency range. Not only do such filters provide high attenuation of harmonics before they can reach the antenna, but arrangements are readily made whereby the harmonic energy of the square wave can be rectified for use as auxiliary dc power. This, of course, further enhances the overall operating efficiency of the transmitter.

Many are not accustomed to think of frequencies of, say, several tens of kilohertz as radio frequency. However, all the attributes we ordinarily associate with RF exist: resonant circuits are used, radiation occurs both intentionally and inadvertently, and high-frequency phenomena, such as dielectric and induction heating, manifest themselves. It should suffice, in this regard, to point out that one of the early uses of wireless communications was to span the Atlantic ocean with frequencies of 12 to 17 kHz. (More modern applications involve very powerful transmitters in about the same frequency region, which provide one-way communications to submerged submarines. The very low frequencies actually comprise a gray area where the line of demarcation between power, audio, supersonic, and radio frequencies is somewhat muddled. For example, a fre-

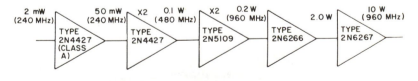

FIGURE 1-19

Typical Amplifier Chain.

Such suggested stage lineups provide the designer and experimenter with relevant guidance at a glance.

(Courtesy of RCA Solid-State Division)

quency of, say, 19 kHz can be shown to possess power, supersonic, and radio-frequency characteristics, all the while displaying circuit attributes familiar enough to the audio buff.

In any event, low frequencies in the 10- to 500-kHz region are commonly found in the following applications:

- Radio transmitters
- Ultrasonic cleaning and mixing
- Ultrasonic welding
- Sonar transmitters
- Switching power supplies
- Inverters
- DC-to-dc converters
- Induction-heating generators and cooking ranges
- Fluorescent lighting supplies

Note that these applications can involve appreciable power levels, from several tens of watts to many kilowatts. (The previously mentioned submarine communicator operates at peak powers in the vicinity of 2 MW.) Those working with the applications listed must deal with such factors as Q of LC circuits, RFI and EMI, harmonic suppression, shield-

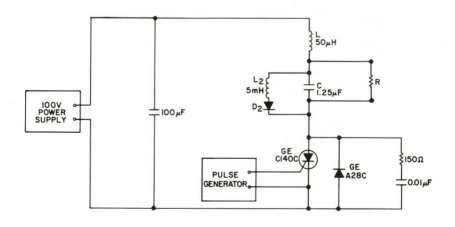

FIGURE 1-20

Basic Inverter Circuit for Producing Low-Frequency Power.
With appropriate selection of SCR and components, several kilowatts of low-frequency (to 30 kHz) power can be developed in load.
(Courtesy of General Electric Co.)

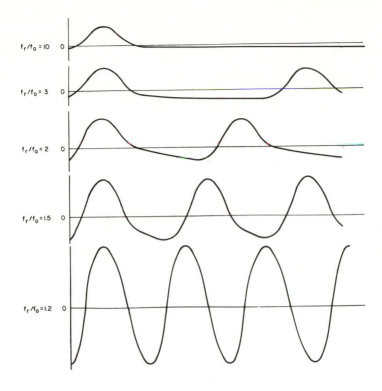

_____ **FIGURE 1-21** _____
**Output Wave Form from Single-SCR Inverter as Function of
Trigger Frequency.**
Sinusoidal wave shape is approached as trigger frequency, f_0,
nears LC resonant frequency, f_r.

ing, shock excitation, parasitic oscillation, frequency stabilization, neutralization, and modulation formats. Thus, one must deal with radio-frequency techniques even if the objective is to produce a dc switching
power supply.

● **Basic SCR Inverter**

The circuit shown in Figure 1-20 exemplifies a simple arrangement
which enjoys widespread use in low-frequency RF applications. The SCR
is used to shock-excite the series LC circuit, but the output frequency is
governed by the trigger pulse rate applied to the SCR's gate. Thus, the
output wave form is not necessarily a continuous sine wave. Figure 1-21
illustrates the output wave form for different trigger rates.

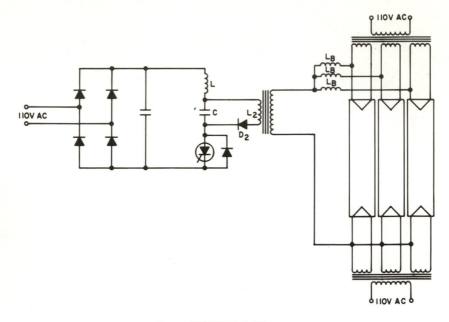

_____ **FIGURE 1-22** _____
Inverter-Driven Fluorescent-Lamp Circuit.
Dimming is accomplished by lowering the frequency of trigger
pulses supplied to gate of SCR.
(Courtesy of General Electric Co.)

In the circuit of Figure 1-20, L_2 serves to help commutate the SCR. Diode D_2 breaks up any resonance that may exist between L_2 and C. (This is most likely to occur at low trigger rates.) L_2 is usually chosen to have ten to about one hundred times the inductance of L. In some circuits, depending upon the trigger rate, and upon the type of SCR, L_2, D_2, or both of these components may be dispensed with. In any event, the SCR used in this circuit is of the *inverter* type or other fabrication specifically intended for high-frequency performance. (In SCR technology, "phase-controlled" types are designed for use in 60- to 400-Hz application. Units intended for operation from several kilohertz to several tens of kilohertz are considered "high-frequency" types and are commonly found in various inverter circuits.)

The output wave forms of Figure 1-21 can be put to direct use in the inverter-driven fluorescent-lamp circuit of Figure 1-22. Here excellent control of the light intensity is obtained by varying the *frequency* of the gate triggers. This is because the voltage pulses delivered to the fluorescent lamps remain high enough for ionization even at very low rates of gate triggers. Inasmuch as the duty cycle of the pulses delivered to the lamps is low under such conditions, the average intensity of light output will be low. There will, however, be no visible flicker because the repetition rate is still far beyond visible perception. L_B designates inductive ballasts. These are needed because of the negative-resistance characteristic of the lamps. They help maintain essentially constant lamp current. Although the nominal operating frequency of this circuit has been in the 5-kHz region, modern SCR's enable frequencies in the 15- to 20-kHz region to be conveniently used.

Three methods of coupling the output of the single-SCR inverter to ultrasonic transducers are shown in Figure 1-23. Ultrasonics applications are important in cleaning, mixing, welding, and in other industrial pro-

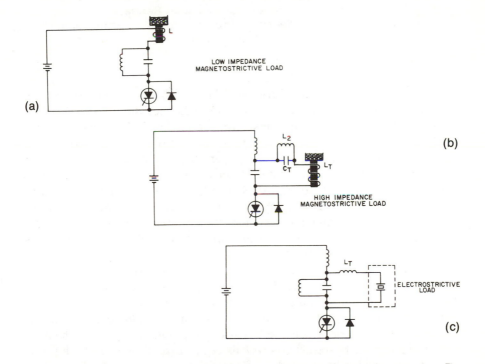

FIGURE 1-23

Methods of Coupling Ultrasonic Transducers to SCR Inverter.
(a) Transducer winding is inductor, L. DC bias is provided for transducer. (b) Capacitor C_T series-resonates L_T. L_2 enables dc bias for transducer winding. (c) L_T series-resonates capacitive reactance of transducer.

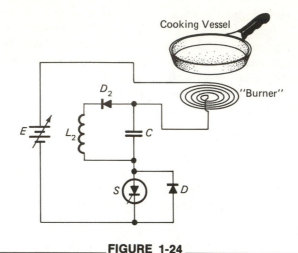

_____FIGURE 1-24_____
Basic Arrangement for Induction Range.
With L in the form of a flat spiral, the cooking utensil is heated by
induced eddy currents.

cesses. Sonar and depth-measuring equipment likewise make use of such
acoustic radiation. SCR's provide good performance up to about 30 kHz.
At higher frequencies, power transistors and power Darlingtons are either
competitive or superior. Some ultrasonic transducers operate in the
megahertz range, where SCR's are no longer suitable. It is interesting to
contemplate that even low-frequency ultrasonic waves in air and in fluids
have wavelengths comparable to high-frequency electromagnetic radiation. That is why such phenomena as cavitation, reflection, refraction,
and focusing into beams are commonly encountered. Thus, even at 15-
kHz or so, the designer or experimenter versed in RF techniques finds
himself in familiar territory.

Yet another application of low frequency RF is the induction range
depicted in Figure 1-24. Here the inductor, L, is formed as a flat spiral so
that optimum coupling can be had with the cooking utensil. Copper ves-
sels are best for this purpose, but aluminum and iron pots can also be accommodated. This basic scheme has also been used to exploit the heating
effect of magnetic hysteresis, as well as eddy currents. Some of the stain-
less-steel alloys might be good candidates for such heat generation. Control of the "burner" is generally brought about by means of a variable
power supply.

A push-pull SCR inverter is shown in Figure 1-25. The gates are trig-
gered by pulses displaced by 180 electrical degrees. The components
shown enable a load power of 1 kW at 13.5 kHz. The basic circuit is very

suitable for frequencies up to the highest permissible from readily available SCR's. This limit has been in the vicinity of 25 or 30 kHz for several years, but 50 kHz may be realizable at reduced power levels from some of the latest high-frequency inverter-type SCR's. The push-pull configuration develops a closer approach to sinusoidal output than the single SCR inverter. And since it does this at lower triggering rates, commutation difficulties at the higher frequencies are more readily overcome. Although the power supply in Figure 1-25 is ac center tapped via the 50 microfarad (μF) capacitors, an even better technique is to use a center-tapped power supply, that is, two series-connected 75-V supplies. The circuit can be made short-circuit proof and current limiting by inserting a 1-μF capacitor in series with the load. The component values can be readily scaled to comply with other than the stipulated frequency.

● Low-Frequency Solid-State Transmitters

Figure 1-26 shows the Continental Electronics type 314E (AN/FRT-89) low-frequency solid-state transmitter. This equipment has a 2-kW power output rating and is intended for operation in the 275- to 530-kHz frequency range. Some of its applications are as follows:

- Ship-to-shore and shore-to-ship communications
- 500-kHz emergency traffic
- Nonidirectional beacon (NDB) navigation-aid transmitter for civil-aviation use

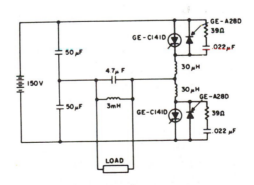

_____ **FIGURE 1-25** _____
Push-Pull Inverter (13.5-kHz, 1000-W).
Sinusoidal wave form is more readily developed by push-pull and
bridge configurations than with single SCR circuits.
(Courtesy of General Electric Co.)

_____**FIGURE 1-26** _____
Two-kilowatt Solid-State Transmitter for 275- 530-kHz Operation.
(Courtesy of Continental Electronics)

- Meteorology information station
- Geophysical and oceanology communication bases
- Automated merchant-vessel reporting program (AMVER)
- Replacement for tube transmitters, especially where greater efficiency and higher reliability are important factors

The block diagram of this transmitter, depicted in Figure 1-27, is interesting in that the push-pull power amplifier is driven from a square-

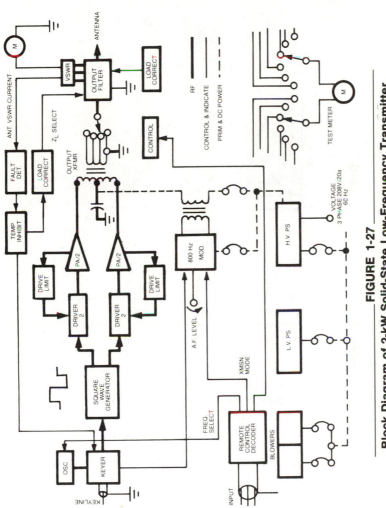

FIGURE 1-27

Block Diagram of 2-kW Solid-State Low-Frequency Transmitter.
The salient feature is class D operation of the push-pull power
amplifier.

(Courtesy of Continental Electronics)

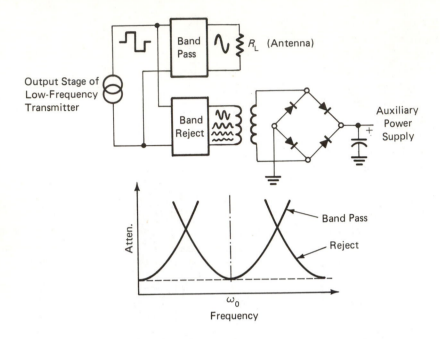

Output Stage of
Low-Frequency
Transmitter

Band Pass

R_L (Antenna)

Band Reject

Auxiliary
Power
Supply

Band Pass

Reject

Atten.

ω_0

Frequency

FIGURE 1-28
**Recapture of Harmonic Energy in Square-Wave Output of
Low-Frequency Class D Power Amplifier.**
The harmonic energy is rectified and filtered to provide auxiliary
dc operating power for the transmitter.

wave source. Such class D operation enables the power amplifier to achieve exceptionally high efficiency by virtue of its essential behavior as a *switch*. Moreover, at even lower frequencies, very similar designs simply substitute SCR's for the power transistors. Thus, below approximately 30 kHz, and where many kilowatts of output power are needed, one would expect to find SCR's. Whether SCR's or transistors are employed as switching devices, the basic idea is that an ideal switch is 100% efficient. The actual efficiency of solid-state switching devices in such low-frequency transmitters is often well above 80%, and with SCR's at very low frequencies, it can be around 90%. (We speak here of the overall efficiency of the final output stage excluding the harmonic filter.)

Radio amateurs and others familiar with class C and class B final amplifiers are aware that these operational modes, too, are described as switching excursions. However, in these instances the transitions from conducting to nonconducting states, and vice versa, are much slower than in the class D mode. Accordingly, the switching losses are considerably greater. Also, power transistors are much more vulnerable to destruction from second-breakdown phenomena when the switching rise and fall times are slow.

Harmonic energy comprises a considerable portion of the square-wave output from a class D power amplifier. Although most of such energy is not dissipated as heat losses in an output harmonic filter, it nonetheless reduces the overall operating efficiency of the transmitter. This is true because energy was consumed from the dc supplies in order to generate and process the square wave form in the first place. The 2-kW low-frequency transmitter utilizes an interesting method for recovering much of this otherwise unused harmonic energy. As shown in Figure 1-28, a band-pass filter is employed to extract the fundamental frequency for delivery to the antenna system. A band-reject filter channels the harmonic energy to a rectifier, and the dc power thereby produced is usefully employed within the transmitter itself. (A similar scheme can be devised with low-pass and high-pass filters.) In any event, it is much easier to attain good filter performance at low RF frequencies: inductors and capacitors tend to behave in more ideal fashion and less trouble is encountered with harmonics and spurious frequencies bypassing the filter networks. For example, in the 2-kW solid-state transmitter described, harmonic radiation is on the order of 70 dB below full-power output into the antenna.

2

The Bipolar Transistor in RF Power Applications

This chapter deals with the unique features of bipolar RF power transistors and their circuitry requirements. It is shown that this is a specialized field, and not merely a frequency extension of previous solid-state power systems. Emphasis is placed upon the selection of use-optimized, rather than "universal," transistors. As opposed to the first chapter, this chapter emphasizes the *differences* between tube and transistor approaches to RF power.

● What Is Unique About RF Power Transistors?

For a long time, the designer and experimenter working with RF power oscillators and amplifiers selected those power transistors with as high a frequency response as was consistent with other parameters, such as voltage and current capability. Generally, this was accomplished by noting the value of f_T in the specification sheet. f_T, known as the gain-bandwidth product, is that frequency at which the common-emitter current gain has fallen to unity. Naturally, a power transistor with f_T much higher than the operating frequency is likely to comply with at least one requisite for RF performance. Later when special switching-type power transistors were developed, one could use the rise and fall times as criteria of satisfactory RF service; a fast "switcher" could be counted on to perform better at high radio frequencies than a sluggish one.

Although it was evident that many tens of watts could be handled at least to the vicinity of several hundred megahertz, the design and operation of a transistor RF power amplifier remained as much an art as a science. A good measure of success was achieved if caution was exercised during tune-up and if the VSWR of the antenna stayed put at a low value. In contrast to tube equipment, there was little margin for even momentary deviation from optimum tuning and adjustment. The counterpart of the benign glow of a tube's anode was invariably a blown power transistor. Also, the internal feedback capacitance of these transistors, being high, made some form of neutralization mandatory. This was not objectional per se, but inadvertent oscillation often sufficed to blow the transistor. And such neutralization tended to be critical and often unreliable.

When ordinary and switching power transistors began to be characterized by *safe operating area* curves, it was thought that RF operation, too, could thereby be made safe. Although one did better by complying with the operating boundaries stipulated by these graphs, it soon became evident that additional "magical" derating had to be applied for RF service. Obviously, what was needed was a power transistor *specifically designed and specified for RF service*. This was forthcoming in the form of a device in which special attention was focused on reducing internal capacitances, preventing hot-spotting (a catastrophic punch-through mechanism similar to secondary breakdown, but accompanying RF operation peculiarly), and special consideration for thermal paths. Also, certain trade-offs were found beneficial. For example, the low collector saturation voltage of the switching transistor could be traded for extended voltage rating.

A unique aspect of bipolar RF power transistors is that operation usually takes place in the attenuation region of the common-emitter current gain curve. This is shown in a qualitative way in Figure 2-1. One can readily see why such transistor amplifiers are prone to low frequency *parasitics*, oscillation and instabilities at frequencies *lower* than the operating frequency. In light of this, it is natural to consider common-base circuits. Indeed, this configuration, with its extended flat-gain characteristic, is much used. However, it too has its shortcomings. The very low input impedance and very high output impedance often make for awkward impedance-matching problems. Moreover, the common-base circuit is also vulnerable to instability, oscillation at the operating frequency. (RF transistors tend to have very low inverse feedback capacitance from collector to base. This enables their use in unneutralized common-emitter circuits. But very small stray capacitance between collector and *emitter*

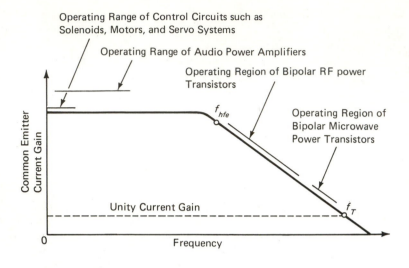

Operating Range of Control Circuits such as
Solenoids, Motors, and Servo Systems

Operating Range of Audio Power Amplifiers

Operating Region of Bipolar RF power
Transistors

f_{hfe}

Operating Region of
Bipolar Microwave
Power Transistors

Common Emitter
Current Gain

Unity Current Gain

f_T

0

Frequency

·FIGURE 2-1·

Relative Operating Regions for Bipolar Power Transistors.
Ideally, it would be desirable to operate RF and microwave
transistors in the flat region of the gain curve. Practically, this is
not feasible in the frequency range where most RF power
systems operate.

can cause oscillation in the common-base connection.) As might be
gleaned from these facts, the common-base configuration is often encountered in power oscillators, particularly in the UHF and microwave regions.

RF transistors are designed and packaged to have very low emitter
lead inductance. This is yet another reason these transistors can be successfully used as common-emitter amplifiers without neutralization.
There are other sophistications incorporated in RF power transistors. For
example, the inductance, resistance, and capacitance of the base terminal
or lead may be manipulated to behave as a low-Q L network. This greatly
facilitates the design of broadband amplifiers and tends to make input impedance matching less critical. Packaging provisions are also provided for
accommodations to stripline tank circuits. This is very important in making UHF applications predictable and reproducible. The tailoring of stray
parameters associated with the leads and the package is an art in itself,
and can yield profitable returns when operation is at UHF and microwave
frequencies. Table 2-1 provides a dramatic illustration of this by comparing the operation of the same chip in four packaging arrangements. Note
that the coaxial package even excels the stripline package. (RF transistor
packages are shown in Figure 2-2.)

● Gain Leveling Provisions

When an amplifier is designed for broadband operation or is intended to be used at different frequencies, it is naturally desirable that its power gain be reasonably constant. This may be desirable in any event for the sake of stability. As already shown, RF gain of a bipolar RF power transistor increases fairly rapidly with decreasing frequency. If matters were left alone, a serious problem could present itself with respect to proper drive power. For example, the drive power level producing optimum output and/or efficiency at 14 MHz could greatly overdrive the transistor at 3.5 MHz. This could result in unnecessary temperature rise in the power transistor and could adversely effect operating bias and linearity. Moreover, in some situations it could provoke problems in the driver stage. Yet another effect is the forcing of the power amplifier into regions of instability.

There are two commonly used circuit techniques for gain leveling the power amplifier so that it will display a near-constant power gain over a desired frequency range. One makes use of a high-pass filter network in the input circuit, such as illustrated in Figure 2-3. Such a two-element filter network, in conjunction with its load resistance, R, can nearly exactly compensate the 6-dB octave slope of the transistor gain characteristic. This is because the filter network, like the "internal low-pass filter" we can postulate in the transistor, operates in its attenuation region. With both filters thus providing *opposite* frequency-response slopes, the net effect is a nearly constant amplitude versus frequency characteristic at the output of the transistor. The load resistance, R, absorbs surplus power in inverse proportion to the frequency so that less base drive is applied to the power transistor as the frequency is lowered. At the same time, the driver stage is not disturbed because the power demanded from it is relatively constant regardless of frequency.

TABLE 2-1

RF Performance Variations with Same Transistor Chip in Different Packages

	f-GHz	P_{in}-W	P_o-W	P.G.-dB	η_C(28V)-%
TO-39	1	0.3	1	5	35
HF-19	1	0.3	1.5	7	45
HF-11	1	0.3	2.2	8.6	50
HF-11	2	0.3	1	5	35

(Courtesy of RCA Solid-State Division)

HF-21
Hermetic
Ceramic-Metal
Coaxial Package,
Large
(JEDEC TO-201AA)

HF-11
Coaxial Package, Small
(JEDEC TO-215AA)

HF-28
Hermetic
Ceramic-Metal
Stripline Package
Grounded emitter
or base

HF-33
Isolated
Electrodes

HF-32
Hermetic
Stripline Package

JEDEC TO-72

HF-31
Hermetic
Ceramic-Metal
Stripline Package
(Studless JEDEC TO-216AA)

HF-19
Hermetic Strip-Line Type
Ceramic-to-Metal Package
with Isolated Electrodes
(JEDEC TO-216AA)

JEDEC TO-39

JEDEC TO-60

HF-12
Molded Silicone-Plastic Case
(JEDEC TO-217AA)

_____ **FIGURE 2-2** _____
Examples of Specialized Packaging and Mounting Provisions Used in RF Transistors.
Power gain, stability, and broadband performance are greatly
influenced by parasitic inductance and capacitance of the
packaging format. The TO-60 package is available with either
grounded or isolated emitter lead.
(Courtesy of RCA Solid-State Division)

Another gain-leveling technique employs *LCR* circuits in a negative feedback path shunted around the power amplifier. In Figure 2-4, gain is stabilized with respect to frequency because the negative feedback increases with reduction of frequency. This method tends to be somewhat less wasteful of power than the aforementioned technique. Such a feedback approach can, if properly deployed, prove beneficial in discouraging instability, one of the ever-present problems in bipolar transistor RF amplifiers. Note, however, that gain leveling is only effective on one side of the series circuit resonance curve. If necessary, reduction of amplifier gain at frequencies below f_r can be provided by other means, such as capacitor C_{in}.

● Basic Transistor Class C Amplifier

The circuit shown in Figure 2-5 is deliberately made similar to that of the basic tube class C RF amplifier of Figure 1-5. Fortunately, this involves little compromise, for both the tube and transistor operate in much the same manner when performing this function. Nonetheless, certain

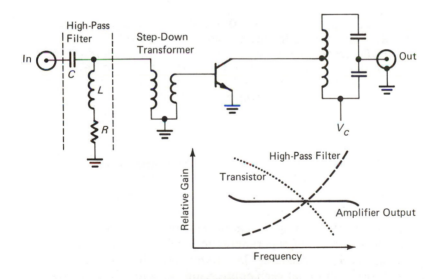

_____ **FIGURE 2-3** _____

Gain Leveling by Means of Input-Circuit Frequency Discrimination.

The input network, *CLR,* is a modified high-pass filter with *R* selected to produce compensation of the transistor's 6 dB/octave roll-off rate. The overall result is a nearly flat frequency response for the amplifier.

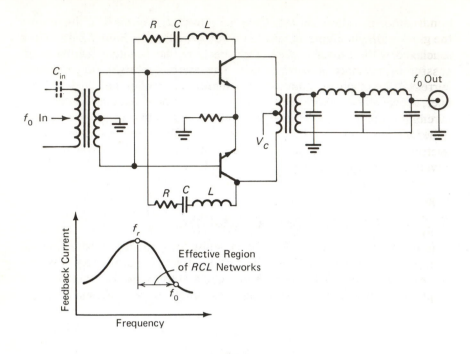

_____ **FIGURE 2-4** _____
Gain Leveling by Means of Negative-Feedback Networks.
The *RCL* networks form low *Q* series-resonant circuits which allow
more negative feedback at low than at high frequencies. The
effectiveness of this method is limited to the frequency region
between f_o and f_r.

practical differences will be observed in the transistor circuit. First, the
transistor is not a forgiving device. In the face of abusive operation, it will
go into thermal runaway, die from secondary breakdown, or be dam-
maged from some other punch-through mechanism. And one does not
have the option of watching for a glowing anode. Such vulnerability tends
to limit one's freedom of experimentation and even requires caution during ordinary tuning procedures. Moreover, the failure mode is usually of
the worst sort, a short between emitter and collector. Thus, an overtaxed
transistor amplifier can endanger the power supply unless suitable protection is provided. (The most general failure mode of tubes, diminished
emission, usually is of benign consequence.)

When a transistor class C amplifier is placed in operation, the con-
structor accustomed to the behavior of tube circuits may feel lost, for the
time-honored manifestations of RF energy may not be present. Even if a

hundred watts of power is available, its presence may not be indicated by the glow of a neon lamp. Nor will a nice RF arc likely be drawn by a pencil touched to a "hot" point (see Figure 2-6). The reason for such apparently passive performance is that the transistor operates at low voltage and high current rather than the converse situation, which occurs in tubes. However, the RF power can readily be proved to be available by suitable measurements conducted at relatively low impedance levels. And though sparking and arcing may be conspicuous by their absence, heating of inductors, capacitors, and leads may be very much in evidence from the relatively high RF currents.

Another difference from tube amplifiers may be observed in push-pull stages. The push-pull tube amplifier often does an effective job of reducing the output of even-order harmonics. Its transistor counterpart is not likely to do so unless a special effort is made to use matched transistors. This is because transistor parameter tolerances are much more sloppy than those of tubes.

Although high-frequency parasitic oscillations can plague transistor amplifiers in much the same manner as tubes, circuit instabilities are more

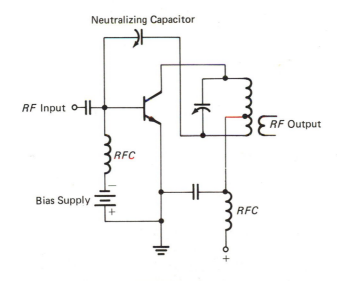

FIGURE 2-5

Basic Arrangement for Demonstrating Transistor Class C Amplifier.

In practical applications, it is often feasible to dispense with neutralization, and it is not usually necessary to provide fixed bias. Also, one or more of the RF chokes can generally be omitted.

likely to occur at relatively low frequencies. This is because transistors are often operated at frequencies where their gain is considerably less than at lower frequencies. This is why resonances of RF chokes used in the input and output circuits often give rise to mysterious oscillations in transistor class C amplifiers.

● **Matching Networks**

Perhaps it may come as a surprise to learn that success with solid-state RF power amplifiers is more intimately bound up with the proper use of matching networks, that is, tank circuits, than any other aspect of electronic circuitry. Although skillful design of the *LC* circuits is also important for optimization of tube amplifiers, it turns out in practice that it is easier to satisfy the requirements of a tube operating at, say 600 V and 100 milliamperes (mA) than those of a transistor consuming 3 A from a 20-V source. In the former case, it is likely that a combination of coils and resonating capacitors can be found in the experimenter's junk box which can produce fair results, or at least point out the path for improvement. One is, however, not so likely to come up with a heavy current-carrying inductor and the relatively large capacitors suitable for the transistor amplifier. The transistor is not as tolerant of departure from proper design of the resonant tanks as is the tube. One reason for this is that transistors are often operated in the region where current gain is appreciably less than at lower frequencies. With an improper output circuit, the transistor tends to be prone to low-frequency instability, an affliction encountered with some tubes too, but with much less severity.

Actually, three important tasks are assigned to the output network or tank. First, energy storage must be sufficient so that a good sine wave is developed at the operating frequency. Otherwise, class B or class C operation cannot obtain. A second and related function is the attenuation of harmonics. This feature is worthy of separate mention because a sine wave that appears good to the eye can still contain appreciable harmonic energy. The third function of the output network is as an impedance transformer so that power may be efficiently coupled from the transistor to the antenna. In actual practice, a fourth property may be cited, the ability to tune out a *small* amount of reactance seen at the antenna feedline.

A number of contradictions, assumptions, and compromises enter the picture. For example, a network should have a high operating Q in order to attenuate harmonics. However, high Q brings with it high circulating currents, and these produce intolerable dissipative losses by the time the operating Q is allowed to exceed 20 or so. And if broadband characteristics of the tuned circuit are desired, it is often quite difficult to achieve a sufficiently high operating Q. Further complicating matters is the awk-

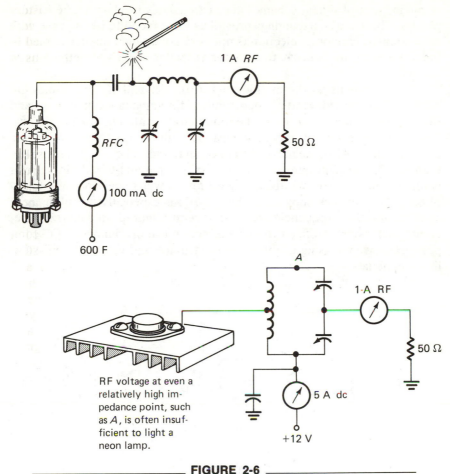

FIGURE 2-6
"Hot" and "Cold" RF Power.
Despite the more spectacular manifestations of RF power in the tube circuit, both RF power amplifiers deliver the same power to the load.

ward fact that the operating Q and the impedance transformation values are interdependent. Finally, at VHF and UHF, the output impedance of transistors varies considerably over the collector voltage cycle.

● Parallel-Tuned Output Network

The simplest output circuit for a transistor RF amplifier is a parallel-resonant LC circuit with either capacitor coupling to the load or inductive coupling via a link. These are shown in Figure 2-7(a), (b), and (c). We are reminded here of similar schemes once extensively used with tube RF amplifiers. The same shortcomings prevail as were found to be the case with tube circuits: harmonic rejection is not very good, and impedance matching to the load, particularly to an antenna feeder, leaves something to be desired.

The output networks on Figure 2-7(d), (e), and (f) are somewhat better. They provide better opportunities for impedance matching and thereby also make better use of the tuned tank for attenuating harmonics. These networks are not, however, as good in these respects as are pi and pi-L networks. Also, tee networks are often found to be better performers than parallel resonant tanks. Nonetheless, the parallel-tuned tank circuit is often found eminently satisfactory for low power stages and output amplifiers. The network shown in Figure 2-7(f) is especially useful in these applications. For frequencies beyond several hundred Megahertz, it becomes increasingly difficult to make effective use of "lumped" LC components, and increasing use is made of stripline and other transmission-line techniques.

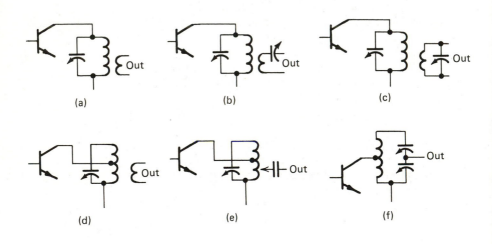

FIGURE 2-7

Various Versions of the Parallel-Resonant Output Tank Used With Transistors.

Networks (a), (b), and (c) tend to find best use with amplifiers operating at power levels up to 1 W or so. At higher power levels, the networks shown in (d), (e), and (f) enable better impedance matches to be made to the collector circuit of the transistor.

● How Transistor and Tube Differences Affect Tuned Circuits

A basic difference between tube and transistor RF amplifiers is in the impedance level of the output circuits. This greatly affects the design of the output tank circuit. To appreciate the gist of this, it is well to contemplate that if a tube were available which operated as a 50-W RF amplifier, drawing 3 A from a 20-V supply, its output impedance level would be virtually the *same* as that of a power transistor operating under such conditions. Specifically, both devices in a class C circuit would look like an RF source with an impedance of $(20)^2/50 \times 2$, or 4.0 ohms. The significant point here is that, both the hypothetical thermionic device and the semiconductor device behave *similarly* insofar as concerns the demands made on the output tank circuit. This causes us to realize that the extremely low output impedance of transistor RF amplifiers is *not* mysteriously related to the fact that the device uses semiconductor material, but is entirely a function of the operating voltage and current. Further confirmation of this stems from the nature of low-power transistor RF amplifiers; they exhibit tubelike output impedances. For example, a transistor RF amplifier operating at 50 V and ½-W output looks like a 2500-ohm source. Interestingly, a tube amplifier with 1-kW output power can also present 2500-ohm impedance to its output tank circuit. Thus the powerful tube amplifier and the flea-power transistor stage might use similar output networks (except, of course, for voltage and current ratings).

Whereas with tubes, it often suffices to merely "apply" the RF excitation to the amplifier input (grid or cathode, as the case may be), with transistors it is usually desirable to use an input-matching network. At higher frequencies this becomes increasingly mandatory.

There is yet another consideration which applies uniquely to transistor RF amplifiers. Even though, as previously mentioned, the transistor may not operate at as high a current gain as it could develop at lower frequencies, the transconductance of power transistors is in an entirely different league from that of tubes. Instead of 10,000 microsiemens (μS) (a "hot" tube), the transistor may operate at several siemens (S) or more. Because of this, it may not be wise to enable the transistor amplifier to develop too much power gain because it will then be difficult to maintain stability. One way to hold down the gain is to use an output network which does not display too high an impedance at the operating frequency. Such a restraint often aggravates the already present conflicts between the network parameters.

So much for the disadvantages. There are also some favorable features pertaining to matching networks for transistors. It is often feasible to use ferrite toroidal inductors with transistors. These greatly reduce feedback problems and are conducive to compact packaging. At higher frequencies, stripline tank circuits circumvent awkward design situations in which tuning reactances and stray reactances become comparable. The stripline technique leads to predictable and reproducible performance at frequencies where lump inductors and capacitors are no longer practical. Inasmuch as transistor RF power is generated at high current and low voltage, less trouble with arcing and flashovers will be experienced. Finally, it is better to have the RF produced in a device of small dimensions so that short connecting leads can be used with the resonating networks. The ability to thus confine RF to where it belongs alleviates at least some problems involving parasitics, stability, and RFI.

● Output Circuits for Medium- and High-Power Amplifiers

Transistor amplifiers designed to deliver more than about 15 W of RF power to the usual 50-ohm antenna feedline generally do not use either the parallel tuned output tanks or pi networks which have been popular with tube amplifiers. An attempt to employ these *LC* circuits would be frustrated by the inability to achieve an impedance match, or by the requirement for impractically large inductors and/or capacitors. Generally, we are faced with the necessity to *step up* the low output impedance of the power transistor to the nominally 50-ohm antenna feedline impedance. It should be recognized that this is just the converse requirement of tube amplifiers, where the tube may typically present an output impedance ranging from several hundred to several thousand ohms. As previously mentioned, the difference in output impedances between the two devices stems from their different voltage-current formats: tubes operate at high voltages and relatively low currents; the opposite is true for transistors.

One may obtain practical insight into this situation by experimenting with various operating voltages and currents of tubes and transistors. This is facilitated by the fact that the approximate formula for output impedance is the same for both devices. It is as follows:

$$Z_L = E_{dc}^2/2P_o$$

where Z_L = output impedance in ohms
E_{dc} = dc voltage applied to the plate or collector
P_0 = expected output power in watts

This equation is applicable to class C amplifiers. It can readily be seen that an RF power transistor intended to produce 15 W of output power when operated from a 12-V supply would present an output impedance of 4.8 ohms. If 45 W of RF output power were forthcoming from such a stage, the output impedance would be a mere 1.6 ohms. Such transistor power stages, as well as higher-powered ones, generally must use basic

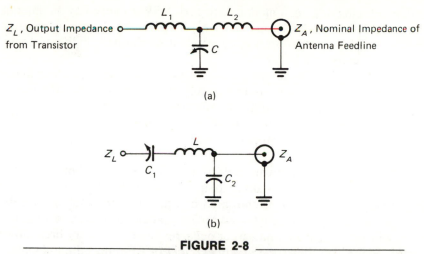

FIGURE 2-8

Basic Output Networks for Transistor Amplifiers Exceeding "Flea-Power" Levels.
(a) T network. (b) L network.

output networks such as depicted in Figure 2-8(a) and (b). It is true, of course, that the several hundred ohm range of output impedance can be obtained at power output levels exceeding 15 W by using RF power transistors which operate at higher than the 12 V of our examples. Then one can conceivably use some version of the parallel-tuned output tank or the pi-network. However, a limit is imposed on how far we can go in this direction by the unfortunate fact that frequency capability and dc operating voltage are trade-offs in the design of power transistors.

● **Survival of Transistors in RF Service**

The mysterious demise of power transistors used in switching applications plagued designers a long time before the phenomenon of second breakdown was discovered and properly investigated. From this experience came the safe operating area (SOA) curves. By ascertaining that the dynamic load line of the transistor never penetrated the off-limit boundaries of the SOA, the problem was solved and its "mystery" gave way to a reliable design procedure. It happened that many power transistors which were fast "switchers" also possessed *some* attributes of a good RF device. In particular, the gain-bandwidth product tended to be high; that is, the frequency response was suitable for at least some RF work. It was found, however, that such transistors were quite vulnerable to catastrophic destruction, this being readily brought about by a departure from optimum tuning or from a moderate or high VSWR caused by an antenna or feeder defect. To make matters worse, there was a pronounced tendency for gradual deterioration even when operating conditions seemingly were optimized.

Empirically, it was found necessary to operate with a safety factor greater than would be conferred by merely paying heed to the SOA curve. Investigation with an infrared "microscope" revealed that the average temperatures pertaining to the junctions, or to the case, were the culprits: the *concept* of average temperature was not valid for RF operation. Instead, there were hot spots in tiny regions *within* the area of the chip. Inasmuch as second breakdown, itself, is actually a thermal phenomenon, it can be appreciated that localized pinpoints of higher temperature than that averaged for the entire chip area can manifest themselves in a destructive manner. Although not completely analogous, from a practical standpoint it is much as if an earlier occurrence of secondary breakdown occurs. Figure 2-9 shows the practical effect of RF-induced hot-spotting. The allowable safe operating area is curtailed. Fortunately, however, it is feasible to construct RF power transistors so that the hot-spotting tendency is greatly diminished.

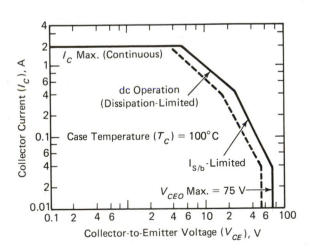

FIGURE 2-9

General Effect of Hot-Spotting on SOA.

The dashed line shows additional operating limitations which must be imposed on non-RF specified boundaries.

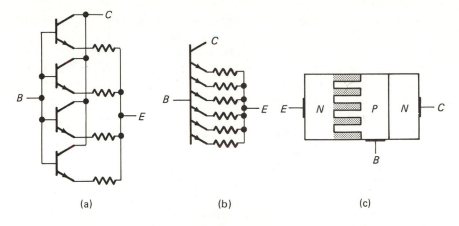

(a) (b) (c)

--- **FIGURE 2-10** ---

Techniques for Reducing Current Hogging and Hot-Spotting in Power Transistors.

The same basic concept applies for all three situations. (a) The use of emitter ballast resistances in dc or audio parallel circuit. (b) Hypothetical structure in which currents in multiple emitters are balanced. (c) Interdigital RF transistor showing emitter geometry. Shaded emitter "fingers" are processed to have higher resistivity than in noninterdigital transistors.

● **Fabrication Techniques for Reducing Hot-Spotting in RF Power Transistors**

The destructive phenomenon of hot-spotting is suggestive of the somewhat more readily identifiable malfunction that tends to occur when we connect two or more power transistors in parallel in a dc or audio circuit. Even if the transistors are fairly well matched in characteristics, it generally happens that the partitioning of current through them will be *unequal*. Not only does this defeat the basic philosophy of dividing the load power equally among them, but as temperatures rise after initial turn on, the inequality tends to become *more* pronounced, generally with one transistor assuming the major burden of load current. Such current "hogging" is inherently *regenerative*, because the transistor that begins to hog the current finds itself increasingly provoked to do so as more current is accompanied by increased heat generation. Such cumulative current hogging, or thermal runaway, is likely to terminate in the catastrophic destruction of the "hungry" transistor.

The remedy to such malperformance is well known, and is depicted in Figure 2-10(a). It consists of inserting *ballast* resistances in each emitter connection to the paralleled transistors. The balancing effect of these ballast resistances is straightforward: if any transistor attempts to hog current, the current flowing through its ballast resistance and emitter circuit develops a voltage drop so polarized as to decrease the forward bias of its emitter-base circuit. It is easy to see that, although various biases will exist among the paralleled transistors, they will incline towards equality of load current division. In most practical applications, one must make a compromise between the value of the resistances and the *quality* of current-sharing attained. Higher resistance results in better current division but also reduces the overall efficiency and voltage-handling capability of the transistor bank. Also, higher resistance decreases power gain because of degeneration.

To provide additional insight into this technique, a hypothetical transistor with multiple emitters is shown in Figure 2-10(b). Here current hogging by any of the emitters is prevented by the same technique of ballasting. (Such a structure is not a fantasy; multiple-emitter transistors are often used in monolithic IC's.) Without the ballast resistances, the multiple-emitter transistor would suffer from the malady termed hot-spotting when ordinary power transistors are operated at radio frequencies. Although the physics of the two compared situations is not exactly the same, the similarity is, indeed, close enough to enhance understanding.

The RF power transistor illustrated in Figure 2-10(c) has an emitter geometry with *interdigited* configuration. Moreover, each of the emitter "fingers" is processed to have higher resistivity than would be the case in ordinary power transistors. This structure then provides the effect of many emitter sites, each with its own ballast resistance. From a circuit viewpoint, these internal resistances are in *parallel* insofar as their effect on overall degeneration is concerned. Therefore, it is feasible to control the current distribution in the chip without degrading the power gain of the transistor. And because the ratio of emitter periphery to area is high, effective use is made of the emitter section.

● **Bias Considerations**

At first inspection, the base-emitter biasing requirements for bipolar RF transistors pose no unique problems. Class A stages are not widely used where the objective is the efficient processing of power. The biasing circuits of class A amplifiers are well known from audio and dc practice. Class B operation appears to offer no unique aspects, and class C operation automatically ensues from operating the base at emitter potential. This comes about naturally by grounding the RF choke or resistance in the base-return circuit of grounded-emitter amplifier circuits. And if one wishes to operate "deeper" into the class C region, the aforementioned choke or resistance is connected to an appropriate negative voltage source rather than to ground. Why look for nonexistent problems?

It happens that most class B (linear) amplifiers actually are more accurately described as operating in the class AB region. That is, they are slightly forward biased so that a small dc idling current exists in the collector circuit when no RF drive is applied to the amplifier's input. This operational mode pays dividends in the form of considerably improved linearity compared to "pure" class B operation. The idling current is adjusted to be a tiny fraction of maximum collector current at ambient temperature. Under this condition, the base-emitter junction of the transistor is approximately at ambient temperature. If RF excitation is applied to the amplifier input for a time, the junctions will undergo temperature rise because of the inescapable power dissipation within the transistor. When the input excitation is *removed,* the idling current will *not* return to its initially set small value. This is because the bias voltage which sufficed to produce the small idling current when the transistor junctions were at, or near, ambient temperature now produces a heavy collector current. This would be serious enough if it merely upset the quiescent operating point of the amplifier.

Because the base-emitter voltage has a negative coefficient of temperature, a situation quickly develops where collector current develops *more* heat, and in turn the excessive forward bias of the base-emitter junction begets yet *greater* collector current. What we have here, of course, is thermal runaway, which seldom satiates itself with less than the catastrophic destruction of the transistor. This phenomenon leaves us with little doubt that we do, indeed, have a biasing problem. It is immediately apparent that what is needed is some means to automatically lower the forward base-emitter bias in response to junction temperature.

A typical method for accomplishing such bias compensation is illustrated by the power amplifier circuit of Figure 2-11. The principle involved is simple enough: the forward voltage developed across the PN junction of diode $D1$ decreases with temperature and thereby also causes the forward bias applied to the base-emitter circuit of $Q1$ to decrease. Of course, we would like to achieve exact tracking of the compensation so that the idling current in the collector circuit of $Q1$ remains constant over a wide temperature range. This objective can be reasonably approached by experimenting with the method of mounting $D1$ and with resistance $R1$. Sometimes a tiny resistance in the ground lead of $D1$, or a high resistance in parallel with it, exert useful control effects. If $Q1$ is a 50-W transistor, $R1$ may be 1 ohm or so, and $R2$ and $R3$ can be 100-ohm resistances with 20-W power ratings $D1$ should be a high-conductance silicon diode such as a 1N4719. For lower-power amplifiers, the 1N4001 is probably more appropriate. Some RF power transistors, such as the RCA 2N6093,

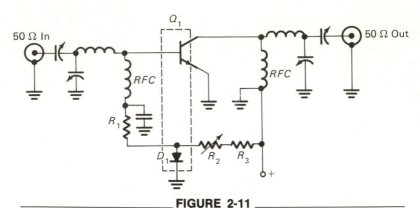

FIGURE 2-11

Basic Diode Compensation Circuit for Preventing Thermal Runaway.

The dashed lines indicate close thermal coupling between the RF power transistor and the diode.

contain an integrally packaged compensation diode. This arrangement provides more intimate thermal contact with the transistor junctions and is conducive to better tracking than is readily forthcoming from an externally mounted diode.

A more elegant scheme for compensating the effect of junction temperature is shown in the biasing arrangement of Figure 2-12. The class AB amplifier involving the RCA 2N6093 RF power transistor develops an output level of 75 W PEP at 30 MHz. The four transistors, Q_1 through Q_4 may be viewed as a current amplifier in the sense that transistor Q_4 is a much more adequate source of biasing current than the simple diode biasing network of Figure 2-11. Circuitwise, these four transistors actually comprise a voltage-regulated bias source with the reference voltage being derived from the temperature-compensating diode and its resistive network. (This diode is integrally packaged with the 2N6093 power transistor.) Unlike the typical Zener diode reference, the voltage from the compensation diode is a variable one. Specifically, it tracks the change of the base-emitter voltage of the 2N6093 with respect to temperature. In this way, the bias voltage and current available from the emitter of Q_4 is controlled over a wide temperature range to cancel the effect of temperature on the quiescent operating point of the RF power stage.

The effectiveness of this biasing scheme in preventing thermal runaway is shown in Figure 2-13. What may not be obvious is that the voltage-regulated bias source actually improves the linearity of the amplifier for single-sideband service. This is because such a ''stiff'' bias source stabilizes the operating point from the effect of the audio-modulated drive signal.

● Basic Problem of Transistor RF Power Circuits

A two-part problem has plagued transistor power amplifiers and oscillators ever since power transistors became commercially available. The maker has had to contend with the frequency capability of the transistor, and the user has had to become versed in the very exacting art of its implementation. The use of such words as "art" may not seem apropos in the face of much published data pertaining to the characteristics of transistors, sophisticated specification sheets, and precise engineering techniques, such as the use of Smith charts and the availability of high-quality RF instrumentation. Yet merely being proficient with the mathematical tools needed will not ensure success. This becomes especially true when

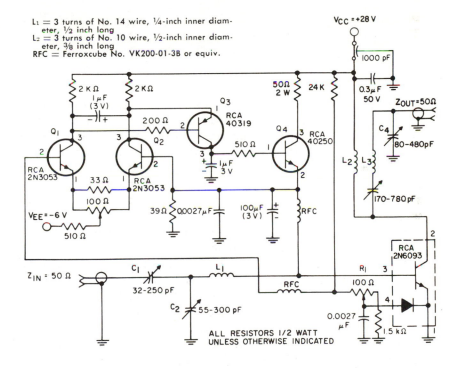

FIGURE 2-12

Bias Compensation Scheme Using Current Amplification to Enhance Performance.

Transistors Q_1 through Q_4 actually comprise a voltage-regulated bias source which senses the voltage developed across the compensating diode.

(Courtesy of RCA Solid-State Division)

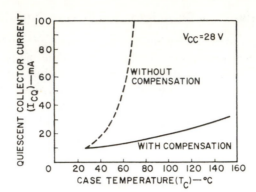

FIGURE 2-13

Effect on Thermal Runaway of the Bias Compensation Scheme of Figure 2-12.
Without bias compensation, thermal runaway would set in as the case temperature of the 2N6093 approached 80°C. Linearity and stability would be adversely affected at even lower case temperatures.

one is working with power levels exceeding several tens of watts and with frequencies of several tens of megahertz and greater. At the microwave frequencies, malperformance readily obtains even at very low powers. It is only natural to ponder the common problem in these applications.

A little investigation reveals the difficulties to stem from low impedance and/or high RF currents. Keep in mind that both input and output impedances of transistors decrease with increasing frequency, and most certainly with power level. If, then, we desire *both* high-frequency operation and a high power level, we must be prepared to use tank circuits and impedance-matching networks with much lower impedances than we might have been accustomed to in tube equipment or in low-level solid-state oscillators and amplifiers. Impedance levels of a fraction of an ohm to several tens of ohms are the order of the day in high-power, high-frequency circuits. At these levels, it indeed becomes an *art* to connect a capacitor to the circuit when the leads of the capacitor may have almost as much inductive reactance as the capacitive reactance being "connected" to the circuit. And those very same leads may have sufficient resistance to spoil the calculated Q of the network or even to heat to fusing temperature from the heavy circulating RF current.

While on the subject of capacitors, it should also be pointed out that the "capacity" can be misleading or meaningless unless *measured at the frequency of use.* The reason is not only because of stray inductance in the leads and structure, but because the dielectric material may behave differently at different regions of the RF spectrum. Other mechanisms such as fringing and radiation also impart a frequency characteristic to real-world capacitors. Capacitance to ground planes and other components is yet another reason why the RF circuit may "see" a different capacity than is stamped on the physical capacitor. Adding further to our difficulties is the problem of ascertaining that a network capacitor is truly connected where it is intended to be. Whereas a one-eighth of an inch one way or the other on a circuit board would be little cause for concern at several megahertz, such sloppy tolerance could cause appreciable detuning at several hundred megahertz.

The traditional way of selecting input or output coupling capacitors or dc blocking capacitors (assuming such capacitors are not actually part of a tuned network) is to merely ascertain that the calculated capacitive reactance is "very low" at the operating frequency. Thus, an 0.01-μF ceramic capacitor might be stipulated for a several megahertz circuit working into a 50-ohm load. According to this philosophy, there would be no adverse effects, other than bulk and cost, if the capacitor were made larger, say 0.1 μF. However, by the time we get into the several tens of megahertz region, and certainly at several hundreds of megahertz, the naivete of such an approach becomes painfully manifest. What has been neglected is the intrinsic inductance, that due to structure and leads, of the capacitor. This makes every capacitor behave as a series-resonant circuit. If a large capacitor is chosen and it turns out that its effective series resonance is near, but not at, the operating frequency, the circuit will not "see" a capacitive component with low impedance, but rather a relatively high impedance path.

From the preceding consideration, it can be seen that a more refined design approach is to select the capacitance no greater than corresponds to series resonance at the frequency of interest. If broadband operation is involved, the frequency of interest is the geometric mean. Thus, if the band extends from, say, 100 to 200 MHz, the mid-band, or frequency of interest, is $\sqrt{100 \times 200} = \sqrt{20,000}$, or approximately 141 MHz. In pursuing this design approach, it is important that the inductive component of the capacitor first be minimized by appropriate selection of capacitor type, then by minimizing (or eliminating) lead length. It is always best to use capacitors made especially for use in the RF spectrum where they will operate. Such capacitors may have short ribbon leads and may be ceramic, mica, or porcelain types. Even better are those with no leads, the capacitor "chips." In schematics, we often see *two* paralleled capacitors. This technique minimizes lead inductance and increases RF current capability.

Bypass capacitors generally are brute-force components of relatively high capacity. They are often electrolytic, high-dielectric-constant ceramic, or Mylar types. In most instances, even when working with low frequencies, these capacitors must themselves be bypassed by one or more *smaller* RF-type capacitors. Here, too, rewarding results may accrue by taking advantage of series resonance.

The effect of low-impedance transistor circuits on inductors is to make them vanishingly small, electrically and physically. Inductance is often simulated by transmission-line elements. In particular, stripline elements are amenable to calculation and result in reproducibility in manufacturing. This is because such elements are lengths of copper strip on PC boards. Their geometric dimensions, together with the dielectric constant and thickness of the board sandwich material, govern the network behavior of these elements. Most important, they facilitate transformation from and to low impedances.

● RF Grounding of the Common Lead

A severe obstacle in dealing with RF power transistors has been the RF grounding of the common lead, the emitter terminal in common-emitter stages and the base terminal in common-base stages. The presence of a fraction of an ohm impedance in the grounding lead can produce pronounced effects in power gain and in stability. This is one reason why RF power transistors have evolved with specialized packaging techniques. For example, the opposed emitter case has not one, but *two* wide terminations for the emitter connection(s). This minimizes the inductance in the emitter grounding path. We are dealing with just a few nanohenrys here, but the inability to achieve a near-zero impedance grounding path was one of the factors delaying the progress of transistor RF power. A little contemplation reveals that impedance in the emitter-ground path produces negative or inverse feedback. And impedance in the base-ground path of a common-base stage produces positive feedback. In actual practice, such ground return impedance tends to produce various erratic results in both amplifier configurations.

In summing up, it is clear that the design and construction of successful transistor RF power circuits is, indeed, an art and a science. It is not to be construed that art of implementation implies that the underlying scientific principles are esoteric or difficult to grasp. Rather, the art comprises an arsenal of practical techniques to be skillfully deployed in an endeavor to comply with basic principles. And the basic principles are generally well reinforced by adequate measurement and performance data supplied by the device manufacturer. On the other hand, some empirical investigation will always be needed because of numerous variables that cannot be accounted for. The purely experimental approach has been left behind, however. For it is no longer necessary to find out by patient and persistent experimentation whether an audio power transistor will magically perform in an RF application.

● Transistor Frequency Multipliers

Familiarity with tube circuits for producing frequency multiplication might lead to the conclusion that the transistor version would simply emulate tube circuitry and design philosophy. That is, a class C amplifier with an output circuit tuned to the desired harmonic would be expected to fulfill this function. This is true, and many frequency multipliers have been implemented in just this way. Sometimes empirical investigation of optimum drive and bias levels has paid worthwhile dividends. However, the approach of merely duplicating tube techniques does not use the bipolar transistor to maximum advantage as a frequency multiplier.

The capacitance variation of the collector-base diode with instantaneous collector voltage is well known. Indeed, this *varactor* phenomenon is responsible for generating more harmonic and intermodulation distortion than would otherwise be the case. Why not use this to advantage in frequency multipliers? Such a question is particularly tantalizing in light of the fact that varactor diode frequency multipliers have long served as efficient frequency multipliers. (By efficient, we imply a relatively low power loss; the passive nature of varactor diodes precludes the possibility of power gain.)

It turns out that *combined* class C amplifier action and collector-base varactor action can, indeed, yield frequency multipliers with greater harmonic producing efficiency than is available from the effects of class C amplifier action alone. To accomplish this, *idler* circuits must be provided for the circulation of harmonic currents. This is quite similar to practice in varactor-diode frequency multipliers. Figure 2-14 depicts the general configurations of common-base frequency multipliers. The series-resonant idler circuits optimize the flow of fundamental and harmonic frequency currents through the transistor. This both enhances varactor action and provides the opportunity for heterodyne products equal in frequency to the desired output harmonic to be generated. With one exception, the remainder of the LC circuitry comprises the conventional impedance-matching networks.

The exception is the parallel-tuned tank circuit appearing in the frequency doubler of Figure 2-14(a). This parallel resonant circuit is tuned to f, the fundamental frequency, and functions as a trap to reduce contamination of the $2f$ output by the fundamental. Note that a $3f$ idler circuit is not needed in the quadrupler. Circuit anomalies of this nature are often learned by empirical investigation.

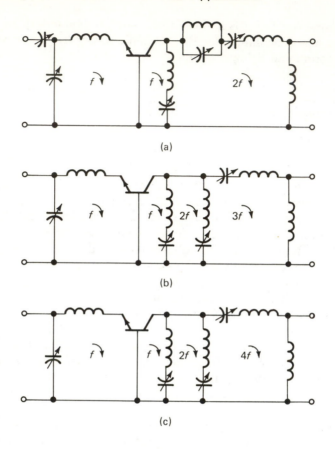

(a)

(b)

(c)

_____ FIGURE 2-14 _____
Common-Base Frequency Multipliers.
(a) Doubler. (b) Tripler. (c) Quadrupler. These simplified circuits do
not show dc supply and bias provisions.

The common-emitter frequency tripler shown in Figure 2-15 also
makes use of collector-base varactor action and is similar to the common-
base circuits. However, the L_2C_2 series-resonant tank in the input is necessary to establish a complete path for $3f$ (450 MHz) current. L_3, L_4, and
C_4 comprise idler circuits for the fundamental frequency (150 MHz) and
the second harmonic. The 450-MHz impedance-matching network in the
output circuit is made up of C_5, L_5, C_6, L_6, and C_7. Some experimentation is generally worthwhile in order to obtain optimum performance.
Generally, compromise is struck between multiplying efficiency and output wave purity. Note that the idler circuits also function as wave traps.

This being the case, it is desirable that these resonant circuits have as high Q as is practical. (Their Q's are, unfortunately, largely governed by the losses encountered in the junctions of the transistor.)

Because the RCA overlay transistors display inordinately low losses from varactor operation, these transistors are eminently well suited for such frequency multiplier service. Some, such as the 2N4012, are actually characterized for this mode of frequency multiplication. This particular transistor can provide 2.5 W at 1000 MHz as a frequency tripler with a collector efficiency of 25%. An interesting aspect of exploiting the varactor action of the transistor is that its frequency capability is approximately doubled.

● Selecting the RF Transistor

The fact that it is of overwhelming importance to select an appropriate device for the particular application at hand is indicative of the progress made in solid-state RF power. Not so long ago, the chief consideration was simply whether amplification or oscillation might occur. And one was prepared to select from a handful of a given type the particular transistor which, for a variety of nebulous reasons, yielded the best performance. Although solid-state RF power continues to be both an art and a science, it is no longer a mysterious art. Enough is known, tabulated, and systematized so that each application can be optimally implemented

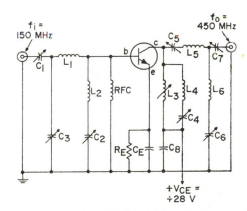

_____ **FIGURE 2-15** _____
Common-Emitter Frequency Tripler.
This circuit requires a series resonant idler tank in the input so
that 3*f* (450 MHz) can complete its path through the collector-base
diode of the transistor.

via a standardized design approach. In selecting appropriate transistors, it should first be recognized that they are manufactured in specific use categories to best satisfy at least one of the following operating features or considerations:

- **Frequency:** One no longer seeks the device with the highest frequency capability, but rather one intended for use in the desired region. Transistors with excessive frequency capability tend to prove fragile when used for low-frequency service.

- **Bandwidth:** While this performance parameter might initially appear to be governed by the associated *LC* networks, this is not entirely true, for the stray and distributed reactances and resistances of the transistor package and leads actually constitute part of the tank circuits. For example, some transistors are so designed that their input impedance represents a low *Q* L-section, thereby facilitating broadband response.

- **Power, current, and voltage:** These parameters are, of course, initially influenced by the circumstances of the application and by cost. It is well to note, at the outset, that one's flexibility is subject to limitations having to do with inherent trade-offs in semiconductor technology. For example, if one is adamant with regard to frequency and power level, it may be necessary to "choose" an *available* operating voltage.

- **Mode of amplifier operation:** Optimized performance features are designed into transistors specifically intended for class A, B, or C operation, as well as for linear, AM, video, frequency multiplication, or pulsed service.

- **Circuit configuration:** Transistors are optimally packaged for either common-emitter or common-base operation. There are also some intended for common-collector operation.

- **General service orientation:** RF power transistors may be marketed for amateur, space, military, marine, aircraft, industrial, or other service. This not only influences the basic operating features, such as frequency capability, power, and voltage, but bears relevantly on cost, reliability, and uniformity of characteristics.

- **Package:** The packaging of an RF power transistor is no trivial matter. In addition to the requirement of low thermal resistance to a heat sink, these devices have unique relationships to their associated circuits. For example, the seemingly tiny inductance of the emitter lead and its RF grounding path exert an almost controlling effect on power gain and operating stability. (Essentially, the same is true for the base lead in common-base circuits.) If one intends to use stripline network elements, the appropriate package style is the SOE terminal arrangement (stripline opposed emitter).

- **Electrical Ruggedness:** In RF power transistors, this figure of merit applies primarily to the ability of the device to stand up under high, or infinite, VSWR conditions, such as might be incurred by a shorted or open antenna feedline.

- **Stability:** Stable operation is very dependent upon the initial design of the transistor and involves such factors as internal feedback capacitance, lead and packaging parameters, spacing and shielding techniques, power gain, and the like. Operational problems can involve both low- and high-frequency parasitic oscillation. Much of the success of producing a good RF power transistor has to do with the ease with which circuit designers can incorporate it in stable amplifiers. The best guide here is to select the transistor supported by the maker's application notes bearing the closest relevance to the intended application. That is, avoid the "general-purpose" approach.

- **Microwave:** Although an extension of the considerations involved under the heading "Frequency" is involved, microwave transistors are unique devices and are often grouped by themselves. In this spectral region, there is more emphasis on common-base circuits, oscillators, and pulsed amplifiers.

The general ideas incorporated in the preceding discussion of operating features are depicted in the typical listings of RF power transistors shown in Figures 2-16 and 2-17.

● Avoiding a Pitfall

The bulk of transistor specifications pertain to operation under *small-signal* conditions. The parameters listed are generally suitable for amplifier circuits biased for the class A mode of operation. However, most RF power amplifiers operate in class C, B, or AB. It turns out that the input and output impedance values of class A amplifiers can be quite different from those in the other operating modes, which are generally classified as *large-signal* values. This is why the design and operation of RF power amplifiers long remained a cut-and-try procedure. To achieve engineering predictability, the measurement of impedances must be made under operating conditions simulating those encountered in actual design implementations.

It would be only natural to ponder the significance of impedance mea-

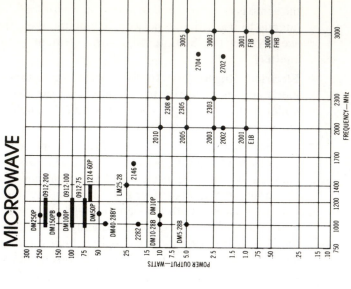

MICROWAVE

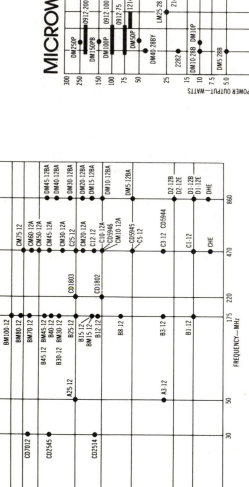

MOBILE

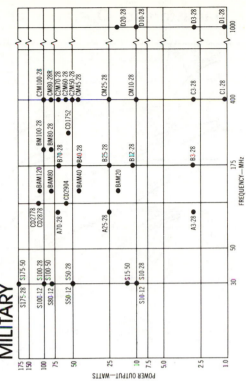

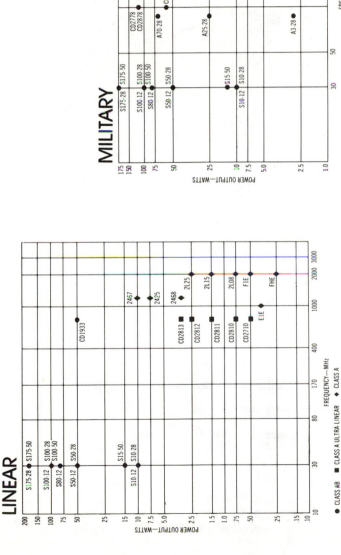

FIGURE 2-16
Typical Listing of RF Power Transistors.
(Courtesy of Communications Transistor Corp.)

71

RF Power Transistors for Operation at 28 V or 50 V

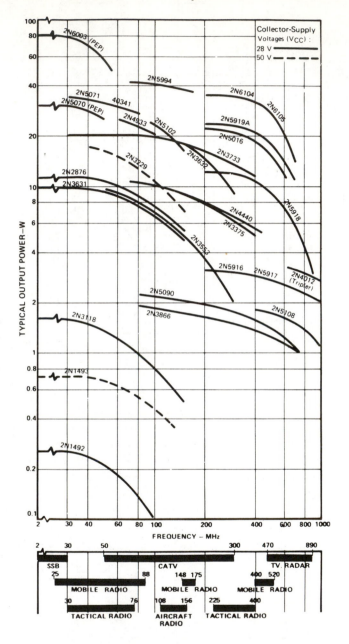

TABLE 2-2

Typical Comparison of Small-Signal (Class A) and Large-Signal (Class C) Parameters

The data pertain to a 2N3948 transistor and show the divergence between commonly published small-signal information and the parameters relevant to most RF power circuitry.

	CLASS A Small-signal amplifier V_{CE} = 15 Vdc; I_c = 80 mA; 300 MHz	CLASS C Power amplifier V_{CE} = 13.6 Vdc; P_o = 1 W
Input resistance	9 Ohms	38 Ohms
Input capacitance or inductance	0.012 μH	21 pF
Transistor output resistance	199 Ohms	92 Ohms
Output capacitance	4.6 pF	5.0 pF
G_{PE}	12.4 dB	8.2 dB

(Courtesy of Motorola Semiconductor Products, Inc.)

surements derived with large signal swings, such as exist in a class C amplifier. For it is clear that any data so recorded can at best be only average values. This is so because a hard-driven transistor behaves as a very nonlinear device. Nonetheless, such impedance values are far more suitable for the design of practical impedance-matching networks than are the small-signal impedance values.

The differences which can occur between small- and large-signal measurement procedures are illustrated in Table 2-2, which compares such impedance data for a 2N3948 transistor. The impedance values depicted are for a parallel circuit. For example, the small-signal input impedance corresponds to a 9-ohm resistance shunted by a 0.012 microhenry (μH) inductance. In contrast, the large-signal input impedance corresponds to a 38-ohm resistance shunted by a 21 picofarad (pF) capacitance. (These values can be readily converted into equivalent series-circuit values by means of the conversion equations given in Chapter 4.)

A word is in order with regard to the class C data. This pertains to the common circuit practice of grounding the base through an RF choke so that both base and emitter are at dc ground potential. As might be suspected, the transistor generally does not operate as deeply into class C as is often the case with tubes. One might say that such a transistor operates in class BC, although such nomenclature is not common. Generally, the same large-signal data are equally applicable to class C, B, and AB amplifiers. The reason is that the class AB amplifier, as used in linear amplifiers, leans much more to the class B mode than class A. And, as we have seen, the class C amplifier is, itself, probably closer to class B than the traditional class C operation of tubes. Even where biasing is such that the conduction period is very short (deep class C), as in certain oscillators and frequency multipliers, the same large-signal impedance parameters produce good results.

3

The Field-Effect
Transistor in
RF Power Applications

The tubelike circuit qualities of field-effect transistors have long merited consideration for RF applications. Hitherto, the power capability of these devices has been so low that their advantages over bipolar transistors were not always compelling. Recent development of the power MOSFET (VMOS power FET) type has radically altered this situation. This chapter develops the fact that these new field-effect devices may well represent a major future trend in solid-state RF power. At the very least, it is shown that these FET's are presently competitive with medium-power bipolar transistors and offer definite circuit and operational advantages.

● Junction Field-Effect Transistor (JFET)

The junction field-effect transistor has been a star performer in RF circuits for many years. Unfortunately, their low power capability has primarily relegated their use to small-signal applications such as RF amplifiers and local oscillators for receivers. In such service, the JFET has acquired a reputation for low noise, high-frequency capability, small and relatively predictable drift characteristics, and minimal loading of resonant circuits connected to its input. Inasmuch as we are interested in solid-state RF *power,* receiver applications will not be discussed. However, in light of the arbitrary definition of power, some allowance must be made for the somewhat nebulous implication of the term. The author tends to look upon those current and voltage levels which require the active device to utilize a heat sink as *power* circuits. In general, this commences at 1- to 3-W output capability. However, the JFET has seen considerable use in applications where outputs of a fraction of a watt to about 1 watt have proved useful. These applications are still far removed from the several tens of milliwatts (or less) level associated with receivers. Moreover, their functions relate to transmitters. Therefore, these low-power FET applications will be considered.

The low-power FET applications involve oscillators, buffers, frequency multipliers, and final amplifiers in amateur QRP transmitters. Another reason that these small devices merit attention is that the fractional-watt power level they develop at VHF and UHF is often as practical as several tens of watts at lower frequencies.

All JFET's are depletion-mode devices; they produce maximum drain current (or nearly so) at zero gate voltage and require application of reverse-bias voltage to reduce the drain current (see Figure 3-1). Both N- and P-channel JFET's are available, sometimes in matched pairs. These

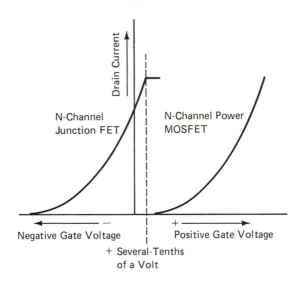

FIGURE 3-1
**Operational Modes of FET's Used for Low- and Medium-Power
RF Functions.**
The junction FET operates in the depletion mode. The power
MOSFET operates in the enhancement mode.

are, respectively, analogous to NPN and PNP bipolar transistors. JFET's are immune to thermal runaway and readily perform well in parallel. However, inasmuch as parameter tolerances are sloppy, at least in "economy" devices, it is a good idea to select reasonably similar units for paralleling.

JFET's tend to make good crystal oscillators, especially where the crystal oscillates in its parallel-resonant mode. Their square-law transfer characteristic can be advantageously used to design efficient frequency-doubling stages. The classic oscillator circuits (Hartley, Colpitts, Clap, Miller, etc.) are all as adaptable to the JFET as to tubes. Indeed, the tube-oriented amateur or designer is likely to feel at home with JFET RF applications. There is one difference, however. There is no counterpart of grid current in FET oscillators or RF amplifiers. The junction diode input section must never be conduction biased. Driving the gate sufficiently positive in an N-channel JFET to cause gate current will limit the RF output in the drain circuit rather than increase it. Otherwise, the JFET is very tube-like. Recall, in this regard, that many, if not most, tubes operate as depletion devices; they require a *reverse* bias to establish an operating point for class A operation. The JFET is sometimes used in class A; examples are VFO's and buffer amplifiers. Where output power and efficiency are important, class C operation predominates (without gate current).

Unless power JFET's are developed for RF applications, it is likely that the now available power MOSFET devices will displace most JFET's. In Figure 3-1 the operational modes of JFET's and power MOS-FET's are shown. (Small-signal MOSFET's often operate in both the depletion and enhancement regions, but are not relevant for our purposes because of their tiny power capabilities.) Note that the JFET may be biased *slightly* positively. This is only on the order of several tenths of a volt, because any higher positive voltage will forward bias the PN junction and thereby limit the operation of the device. It will be recognized that various tubes can be classified as depletion or enhancement devices, with most sharing both modes. (Bipolar transistors operate in the enhancement region of Figure 3-1, although this terminology is not employed to describe their operational mode.) The same applies to zero-bias tubes. The significance of these operating modes pertains primarily to biasing arrangements. The dynamic characteristics of two devices can be quite similar even though their operational mode and bias circuits are different.

Amateurs have been traditionally adept at squeezing more power out of devices than specified by manufacturer's ratings. In the JFET, however, they have not been successful. This has been frustrating because of the other fine qualities of the device. (Overseas technology has developed power JFET's, but not much information is yet available with regard to their performance in RF circuitry.) Figure 3-1 shows why JFET's cannot

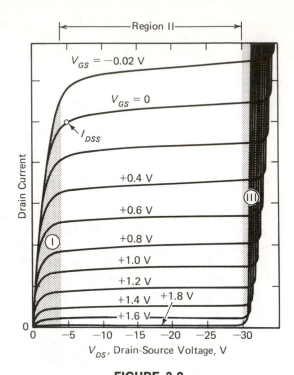

FIGURE 3-2
**Output Characteristics of a P-Channel JFET Showing
Drain-Voltage Limitations.**
Region III represents avalanche breakdown in the gate-source PN
junction.

be overdriven and Figure 3-2 shows that one also runs into difficulties if
an attempt is made to raise the drain voltage. The sudden discontinuity in
the curves results from avalanche breakdown in the gate-source section of
the device because a portion of the drain potential adds to the reverse bias
of the gate. Thus, a larger heat sink, or intermittent service (such as CW),
does not result in increased power output.

● Simple Circuitry at Low-Power Levels

Radio-frequency circuitry using JFET devices tends to be simpler
than its bipolar transistor counterparts. Actually, the JFET is usually
compatible with RF whether it was deliberately designed for such service
or not. One reason is the junctionless output section of the JFET. Another
is the way in which the gate-junction is operated—in its reverse-bias
mode. The net result is that a grounded-gate circuit, such as shown in Fig-

ure 3-3, can perform well over a frequency range of 20 to 200 MHz. The input impedance is actually much higher than 50 ohms and displays virtually no reactive component over this frequency range. Thus, the input network problem, so important with bipolar designs, is of little consequence here. Transformer $T1$ in the output circuit is a transmission-line type which provides a 16 to 1 impedance step down. The beaded leads function as RF chokes at the higher frequencies. (Among the many ferrite beads that are suitable are the Carbonyl J units made by the Pyroferric Company. Generally, high permeability and fairly high loss (to keep the Q low) are the criteria for selection of such ferrite beads.)

A surprising feature of such a circuit as Figure 3-3 is its respectable performance as a *linear amplifier*. This appears to contradict the well-known fact that the large-signal operation of a JFET is anything but linear. However, the nonlinearity of the device follows a nearly true square-law relationship. This does not give rise to the third-order intermodulation distortion by which linear amplifiers are evaluated. Thus, by application of an appropriate bias voltage, V_{GS}, performance such as depicted in Figure 3-4 can be attained.

As already pointed out, the power ratings of most JFET devices have been quite low. A JFET such as the Siliconix U322 is rated for a power dissipation of 3 W with its TO-5 package at 25° C. This is a relatively high-powered unit!

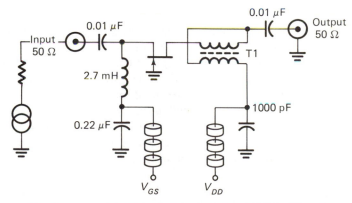

T1—6 turns no. 22 AWG twisted pair wire on 0.375 in. Diameter
Indiana General F625-902 toroid core.

FIGURE 3-3

JFET Building Block Buffer or Low-Power Output Amplifier.
Grounded-gate circuit is noteworthy for its simplicity and its
broadband characteristics.
(Courtesy of Siliconix)

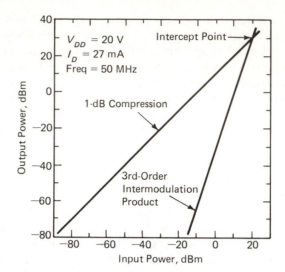

FIGURE 3-4
Linear Amplifier Performance of the Grounded-Gate Amplifier of Figure 3-3.
The output current of the JFET relates to input voltage by a nearly true square law, but *third-order* intermodulation products tend to be low.
(Courtesy of Siliconix)

The reason grounded-gate amplifier circuits are popular with the JFET designers is that stability and flat frequency response are readily forthcoming without the drawback displayed by the bipolar transistor in analogous circuits. This drawback is the very low input impedance which the forward biasing of the emitter-base junction engenders.

All things considered, it is tantalizing that multiwatt JFET's are available for RF purposes. Certain overseas developments suggest the possibility of such devices in the near future. However, the dramatic development of the power MOSFET now makes such a "crash program" less compelling.

● **Power FET's: "Solid-State Tubes"**

A true technological breakthrough has been made with regard to MOS field-effect transistors. These devices are no longer restricted to signal-level power ratings. Although the popular consensus of field-effect transistors as flea-power devices with inherent electrical fragility may, indeed, have had a factual basis, it is also fact that we are now in an era of

FET's with current capabilities of approximately 15 A and with 400 V ratings or more. That some of these devices are of the insulated-gate (MOS) variety with *rugged* electrical characteristics initially taxes our sense of credibility. Then to learn that such FET's can operate efficiently as power amplifiers and oscillators well into the UHF region requires a rethinking of one's attitudes. Not only have MOS field-effect transistors evolved into capable RF power devices, but they display the following advantages over bipolar types:

- They do not suffer from secondary breakdown, there being no PN junction in the output section (i.e., the drain-source circuit). Also, there is no hot-spotting.
- They are not vulnerable to thermal runaway. This feature stems from the positive coefficient of channel resistance with respect to channel current.
- There is no current hogging; units can be directly connected in parallel without ballast resistances.
- Frequency capability is ultimately limited by stray reactance, rather than by charge-storage effects. This makes possible a very high gain-bandwidth without greatly detrimental trade-offs of voltage or power ratings.
- The input impedance can be said to be tube-like. This simplifies certain input tank problems and tends to be less demanding on the drive circuitry.
- Broadbanding is often more readily accomplished.
- Low-frequency instability is less likely because operation generally occurs at or near the maximally available power gain of the device.
- The VSWR of the load does not endanger the device.
- Excellent proportionality exists between input voltage and output current. Thus the device is inherently applicable to linear RF power amplifiers. (It is not always easy to obtain such linearity from bipolar RF power transistors.)
- More reasonable impedance transformation is often possible because of operation at higher voltages and lower currents.
- Cost is at least comparable to ordinary power transistors and may prove to be lower than true RF types of bipolar transistors.
- Biasing tends to be less critical.
- There is no varactor action in the output (drain-source) circuit. This reduces harmonic generation.

● Ultrafast switching performance renders feasible switching-type
RF amplifiers in which theoretical efficiency is 100% (maximum
theoretical efficiency of the ideal class C amplifier is about 85%).

● Other Unique Features of the VMOS Power FET

A curious, but potentially useful characteristic of the VMOS power
FET is its inordinately low noise figure, in the vicinity of 2.5 dB at
150 MHz. To be sure, noise figure does not have much import for the designer of power output stages. However, the low noise figure of this device in conjunction with yet another strange feature suggests novel applications. This feature is its substantially constant input and output
impedances from several microwatts to several tens of watts or more of
output power. This, of course, bodes well for linearity in single-sideband
amplifiers and for frequency stability in oscillators, but these features also
suggest an application which would be almost unthinkable with bipolar
RF transistors.

The great emphasis on *transceivers* during the past decade has taxed
the circuit designer's ingenuity for techniques leading to simplicity and
cost effectiveness. From the preceding paragraph, it is apparent that the
very same VMOS FET RF final amplifier stage in a transceiver could *also*
function as a receiver RF input amplifier. This, of course, requires a
reorientation of attitude; one certainly is not conditioned to think of a bipolar power transistor in such a dual role.

It appears, also, that the VMOS power FET should be ideally suited
for crystal oscillator circuits. The low input impedance of bipolar transistors has always been an adverse factor in such applications. And, other
things being equal, the VMOS power FET should also be a better candidate for attaining frequency stability in VFO's. Extending our reasoning
farther in this general direction, this device also appears to be better
suited for achieving isolation in buffer amplifier stages. (Additionally, its
higher input impedance is less likely to exert a loading effect upon the
VFO or the preceding stage.) Many VFO designs have made use of small-
signal junction FET's to accomplish these various objectives. The low
power-handling capability of these JFET's has often posed obstacles,
however.

Yet another unique feature of VMOS power FET's should be of interest to the far-sighted designer or imaginative experimenter. This pertains
to the fact that the input of these devices directly interfaces with CMOS,
TTL, DTL, and MOS logic families. Thus, digital circuitry could be used
to accomplish switching operations in class D fashion. Under such conditions, an appropriate *LC* tank, or harmonic filter in the output circuit,
would extract the fundamental frequency from the square wave. Because

the rise and fall times are on the order of 4 nanoseconds, (ns), such an approach merits consideration.

It is anticipated that as the technical community outgrows its now obsolete notion that MOS technology is limited to flea-power devices, these and other unique applications will naturally evolve.

● V-Groove Structure of the Power FET's

The difference in fabrication between the new power FET's and the conventional flea-power MOSFET is illustrated in Figure 3-5. The V-groove gate structure shown in part (a) allows the controlled source-drain current to follow a vertical path through the four semiconductor regions. Note that the drain region occupies a large area. This enables the case to heat sink thermal resistance to be low. Because the V configuration involves two channels, the single metallized gate controls two vertical current paths. The n^- region is an epitaxial layer and behaves as a space charge region, enabling much higher drain voltages to be applied than in the case of the ordinary MOSFET structure shown in part (b). Another beneficial effect of the n^- layer is a considerable reduction in output capacitance and in drain-gate feedback capacitance—of prime significance in RF circuits.

The power MOSFET operates in the enhancement mode; that is, output current is cut off when the gate has the same potential as the source. A

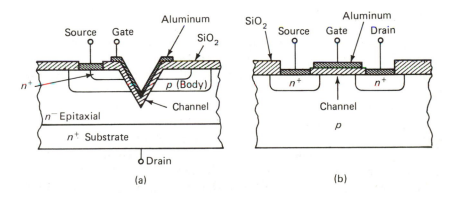

FIGURE 3-5
Comparison of the Power MOSFET Structure with Ordinary MOSFET Geometry.
(a) The V-MOS structure of the new power units. Source-drain current path is vertical through the four semiconductor regions. (b) Previous MOSFET configuration—limited to signal-level power.

positive gate bias on the order of 10 V or so is required for rated channel current in the source-drain circuit. However, unlike either tubes or bipolar power transistors, this positive bias is *not* accompanied by dc bias current. Therefore, the dc input impedance is very high, comparable to that of tubes operating with negative grid bias. A corollary of this is that the power MOSFET requires very little power from its driver stage. There is no input current analogous to grid current in a class C tube amplifier.

Those familiar with low-power insulated-gate transistors are aware of their vulnerability to destruction from static electricity. Merely handling one of these devices often suffices to blow out the gate structure. Many, if not most, of the power MOSFET's have internal monolithically fabricated protection diodes associated with the gate structure. Those that do not require very careful handling. However, once in their circuits, the susceptibility to gate damage from transients or static charges is greatly reduced. Nonetheless, it is unwise to use an ungrounded electric soldering iron around a circuit using the power MOSFET. Those power MOSFETs which are *specifically* intended for RF applications generally do not incorporate internal gate protection in order to avoid the adverse effects protective diodes have on the input impedance.

● Some Revealing Features of the Power MOSFET

Let us suppose that, as a seasoned veteran of bipolar transistor circuits, you encounter a set of operational curves such as those shown in Figure 3-6. Casual inspection might not suggest anything to stimulate unusual interest. Let us, however, take more than a superficial glance at these curves. The output characteristics depicted in Figure 3-6(a) appear familiar enough; they resemble those of pentode tubes and of bipolar transistors. However, one rarely sees such curves for bipolar transistors representing the most often used common-emitter connection. Rather, they resemble those generated in the common-base circuit. (Common emitter curves are less evenly spaced and have appreciable slopes.)

Unfortunately, common-base circuitry often presents us with awkward impedance levels, especially in RF work. But the flat curves of Figure 3-6(a) *are* for the "common-emitter" (now called common-drain) connection of the power MOSFET. The horizontal aspect of these curves indicates very high output resistance. The even spacing of many of them indicates an extensive linear region of operation. We may rightfully infer that the power MOSFET is a good candidate for a class AB linear RF power amplifier.

Figure 3-6(b) reveals similar information in a different way. Here we see that the transconductance of the device is flat over an appreciable operating range. Note that the flat region embraces the evenly spaced operating range of Figure 3-6(a). Note, also, the fact that the transconductance is in the vicinity of 275 mS, a "hot" *tube* indeed! It is true that power bipolar transistors exhibit tens of siemens of transconductance. But the power gain or current gain of bipolar transistors is low or moder-

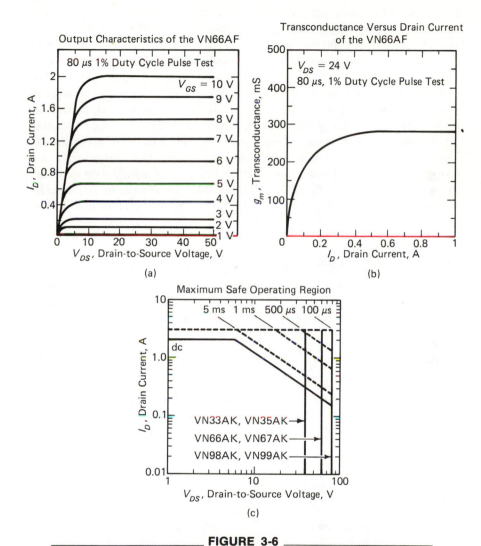

FIGURE 3-6

Curves Illustrating Features of the Power MOSFET.
(a) Pentodelike output characteristic. (b) Constant transconductance over appreciable operating range. (c) No secondary breakdown limitation in SOA curves.
(Courtesy of Siliconix Corp.)

ate because of their current-consuming input circuits. Another way of making such comparisons (not revealed by these curves) is to contemplate the equivalent beta of power MOSFET's, which is on the order of 1 billion! In this respect, the power MOSFET can be likened to a "solid-state tube."

In the safe-operating-area graph of Figure 3-6(c), something is conspicuous by its *absence,* a boundary imposed by secondary breakdown. As already mentioned, the power MOSFET has no such limitation because there are no PN junctions in its output section or channel. Thus, the SOA curves are not unlike those which could be drawn for a simple passive element, such as a resistor. A little thought reveals that even a resistor has a maximum safe current, a maximum safe voltage, and an operating region bounded by power dissipation considerations. The absence of secondary breakdown or of hot-spotting regions is one of the salient features of the power MOSFET. The practical manifestation is that the device is not likely to suffer catastrophic destruction because of an unresonated tank circuit or because of high VSWR presented by an antenna feeder line. By the same token, it will not be vulnerable to excessive modulation or to transients.

● Example of MOS Power Capability

Those accustomed to the milliwatt range of ratings of small-signal MOSFET's, or the fractional-watt operating levels of JFET's, will be surprised by the specifications of the VN84GA power MOSFET shown in Figure 3-7. Although this particular power MOSFET is not optimally designed or packaged for VHF application, its 50-ns switching times suggest good performance in the popular amateur high-frequency bands. Another favorable factor portending satisfactory RF operation is the absence of an input Zener diode. Note the forward transconductance of 2 S; compare this with the "hottest" transmitting tube you can find. It will be evident that the oft-used adjective "tubelike" renders the power MOSFET an injustice.

● Input Circuit of the VMOS Power FET

The VMOS power FET shares with small-signal MOS transistors vulnerability to destruction from static charges applied to its gate. Once the device is in its circuit, this danger may no longer exist, but during initial handling, consideration must be directed to static-generating conditions, such as walking on synthetic fiber carpets, wearing crepe-sole shoes, and the like. Also, it is wise to watch out for leakage currents from soldering irons during installation. Many VMOS power FET's have monolithically built-in Zener diodes in the gate-source circuit and are relatively immune

TO-3

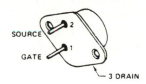

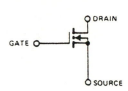

ABSOLUTE MAXIMUM RATINGS

Maximum Drain-Source Voltage .... 80 V
Maximum Drain-Gate Voltage .... 80 V
Maximum Continuous Drain Current .... 12.5 A
Maximum Pulse Drain Current .... 15 A
Maximum Gate-Source Voltage .... ±30 V
Maximum Dissipation at 25°C Case Temperature ..... 80 W
Linear Derating Factor .... 1.56°C/W
Temperature (Operating and Storage) ...... −55 to +150°C
Lead Temperature
    (1/16″ from case for 10 seconds) .... 300°C

ELECTRICAL CHARACTERISTICS (25°C unless otherwise noted)

		Characteristic	Min	Typ	Max	Unit	Test Conditions
1	S T A T I C	BV_{DSS} Drain-Source Breakdown	80			V	$V_{GS} = 0V$, $I_D = 100$ mA
2		$V_{GS(th)}$ Gate Threshold Voltage		2.5			$V_{DS} = V_{GS}$, $I_D = 10$ mA
3		I_{GSS} Gate-Body Leakage			100	nA	$V_{GS} = 10$ V, $V_{DS} = 0$
4		I_{DSS} Zero Gate Voltage Drain Current			1.0	mA	$V_{DS} = $ Max. Rating, $V_{GS} = 0$
5		$I_{D(on)}$ ON-State Drain Current	10			A	$V_{DS} = 25$ V, $V_{GS} = 10$ V (Note 1)
6		$R_{DS(on)}$ Drain-Source ON Resistance		0.3	0.4	Ω	$V_{GS} = 10$ V, $I_D = 10$ A (Note 1)
7	DYNAMIC	g_{fs} Forward Transconductance	1.5	2.0		℧	$V_{DS} = 20$ V, $I_D = 5$ A (Note 1)
8		C_{iss} Input Capacitance		640			$V_{GS} = 0$, $V_{DS} = 25$ V, f = 1.0 MHz
9		C_{rss} Reverse Transfer Capacitance		50		pF	
10		C_{oss} Output Capacitance		300			
11		t_{on} Turn-ON Time		50		ns	$V_{DS} = 20$ V, $I_D = 10$ A (Note 2) $R_L = 2Ω$
12		t_{off} Turn-OFF Time		50			

NOTES: 1. Pulse Test − 300 μs, 1% duty cycle
    2. See switching time test circuit

VNG

FIGURE 3-7
Operating Specifications for the VN84GA Power MOSFET.
The data indicate a respectable power-handling capability.
(Courtesy of Siliconix Corp.)

to damage from static charges and leakage transients from the power line. Unfortunately, this generally applies only to those devices intended for audio, servo, and low-frequency service. Although the protection conferred by the input Zener diode would be welcome in RF applications, the side effects pose difficulties. These include increased power consumption from the driver stage and increased input capacitance. Also, the Zener diode, itself, is vulnerable to burnout.

Although the gate exhibits essentially infinite impedance to a dc controlling signal, this is not the case in RF applications where input capacitance, together with the Miller effect from internal drain-gate feedback, makes the gate a relatively low impedance. However, the impedance at the gate is, nonetheless, at least an order of magnitude greater than would be encountered at the base of a bipolar transistor of similar power capability. This has much practical significance, for many of the impedance-matching problems associated with the inordinately low input impedance of bipolar transistors are averted. At VHF, the VMOS power FET may typically display an input impedance in the several tens of ohms, contrasted to several tenths of an ohm in bipolar power transistors.

In most RF power applications, the VMOS FET is used in the common-source circuit, which is analogous to common-emitter and common-cathode configurations in bipolar transistors and tubes, respectively. Thus, the drive signal is applied to the gate. This results in stable operation from low RF through UHF. The tendency toward low-frequency oscillation experienced in bipolar transistor amplifiers is practically nonexistent with these devices.

Unlike certain other MOS devices, the VMOS power FET operates in the enhancement mode; output (drain) current is zero with no input sig-

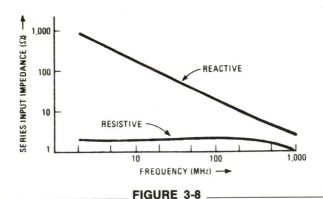

FIGURE 3-8

Common-Source Input Impedance of Typical RF Power MOSFET.
Curves represent a nearly "pure" capacitance in series with a low
and essentially constant resistance.

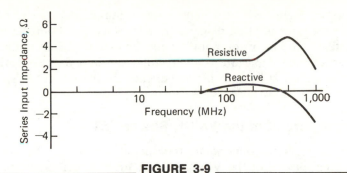

FIGURE 3-9

Common-Gate Input Impedance of Typical RF Power MOSFET.
The flat resistive input impedance portends readily attained
wideband performance up to several tens of megahertz.

nal. Thus, a measure of fail-safe performance is inherent in its very nature. The enhancement mode usually results in simplified bias circuitry and in a more direct approach to class B and class C operation. At the present writing, most development has centered on n-type devices (the counterpart of NPN bipolars). Near-optimum class C performance generally obtains from zero-bias operation (i.e., with the gate at source potential). From the standpoint of circuit simplicity, this is a desirable feature.

Typical input impedance versus frequency for a power MOSFET is shown in Figure 3-8. This is for a series-equivalent circuit; as with a conventional capacitor, the impedance can be represented as either a very high resistance in parallel with a capacitance (parallel-equivalent circuit) or a very low resistance in series with a capacitance. The reactive component will be seen to have a constant slope; that is, it is a straight line. This has a special significance on a log-log plot; it implies that the capacitance does not change with frequency. This contrasts to the frequency-dependent input capacitance of bipolar transistors. A related factor, not indicated on this graph, is that the input capacitance of the power MOSFET does not change with drive level, again contrasted with the situation in bipolar transistors. If we were not otherwise informed, the reactance curve of Figure 3-8 could be construed to represent a high-quality capacitor. Indeed, the combined impedance plots, reactance and resistance, could be so interpreted.

When first encountered, the common-gate input impedance of the power MOSFET often evokes surprise. As seen in Figure 3-9, the input impedance can be purely resistive out to several tens of megahertz. This follows directly from the common-source curves of Figure 3-8, but the relationship requires the mathematical rather than the intuitive approach.

The plots of Figure 3-9 also show why common-gate performance at higher frequencies is often plagued by instability, for both the resistive and reactive components of input impedance then depart from their near-ideal levels. Such departure is likely to cause impedance mismatch in such an amplifier, at best, or high-frequency oscillation at worst.

● Output Circuit of the VMOS Power FET

From a practical standpoint, the salient feature of the output, or drain-source, circuit of the VMOS power FET is that there are no PN junctions. Rather, there is only a relatively simple conductive path. When compared to the bipolar transistor, the implications of this junctionless output circuit are significant. First, the destructive phenomenon of secondary breakdown is absent. Also, because of the lack of PN junctions, there is no varactor effect from voltage-dependent PN capacitance. This results in an easier-to-filter harmonic spectrum in class B and C amplifiers. The output circuit is bilateral in its conductive properties so that no protective diode need be used in class D or class F amplifiers. This is important, for such diodes would tend to degrade the efficiency of these high-frequency switching amplifiers. (This is one of the reasons the bipolar transistor has not been successfully deployed in these otherwise efficient operational modes for RF amplifiers.)

Another closely related aspect of such a junctionless output circuit is the absence of thermal runaway. Whereas in a bipolar transistor excessive junction temperature regeneratively increases output current, thereby increasing temperature further, the very opposite situation prevails in the VMOS power FET. Here excessive temperature, from whatever cause, *reduces* the conductivity of the output circuit, thereby tending to self-correct the overheated operation. In any event, the device does not drive itself to thermal destruction. A corollary of this desirable characteristic is that VMOS FET's can be paralleled without fear of current hogging by one of the units. No ballast resistances are needed in the parallel combination, for the internal mechanism which prevents thermal runaway also mitigates against current hogging. This property is summed up by stating that the device has a negative coefficient of output current with respect to temperature. (In the technical literature, one sometimes finds allusion to the "positive temperature coefficient" as being responsible for absence of thermal runaway. What is meant here is a positive coefficient of *output resistance* with respect to temperature.)

It should not be construed that the absence of thermal runaway stems from the conductive properties of the drain-source silicon element itself. Indeed, silicon increases its conductivity with temperature. What happens in the *overall* device, however, is that the *transconductance* decreases with temperature. This counteracts any tendency of output current to runaway when the FET becomes hot.

The transfer characteristics of the power MOSFET are best described by transconductance, the same parameter that has proved so useful with tubes. Transfer curves such as shown in Figure 3-10 provide much information for the designer. Transconductance is readily obtained as the ratio of change in drain current caused by a small change in gate voltage while drain voltage is held constant. Applying this concept to Figure 3-10, the transconductance in the linear region is readily found to be 250,000 μS.

● Class D RF Amplifiers

The operating modes that have been most common with tube amplifiers have been class B (or AB) and class C. The former have been used to provide linear amplification of single-sideband signals and the latter has proved most useful for CW, AM, and FM modes. The same has been true of transistor amplifiers, except that class A operation is often used at low power levels. With both tubes and transistors, class B amplifiers are sometimes used for CW and FM despite its somewhat lower efficiency

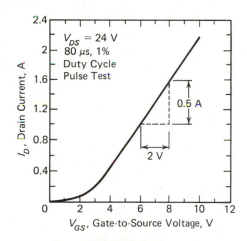

_____ **FIGURE 3-10** _____

Transfer Characteristic of the VMP-4 Power MOSFET.
This curve depicts a threshold of approximately 2 V and very good
linearity for drain currents exceeding 400 μA. The
transconductance is

$$\frac{0.5 \text{ A}}{2.0 \text{ V}} = 250,000 \ \mu s$$

(Courtesy of Siliconix Corp.)

compared to class C operation. Also, the class B (or AB) amplifier, because of its linearity, can be employed to provide power boost for AM. With both tubes and transistors, the class C operating mode has enjoyed universal application in frequency-multiplier stages.

The maximum theoretical efficiency for class A stages is 50%, that for class B operation is 78.5%, and class C amplifiers can attain efficiency percentages around 85%. In all instances, the cited efficiencies prevail only when the amplifiers are driven to maximum output; at lower outputs, efficiency is less.

There is yet another mode of amplifier operation, one which is capable of even greater efficiencies than is ordinarily possible with the class C amplifier. This operating mode, class D, causes the amplifying device to behave as a near-ideal switch. Such a switch is on half the time and off the remaining half. To perform this way, the transition time between switching states must approach zero. The salient feature of an ideal switch is that there is no power loss when it is on because the on resistance is construed to be zero. There is no power loss when it is off because its off resistance is construed to be infinite. And if the switching transition time is zero, no power loss can occur between the states. Of course, the output of such a switch is a square wave and must be adequately filtered so that the load is presented with a clean sine wave. Such filtering can be best accomplished with the aid of a harmonic filter inserted between amplifier and load.

Although it is commonly said that the theoretical maximum efficiency of the class D amplifier is 100%, this statement is not quite applicable to

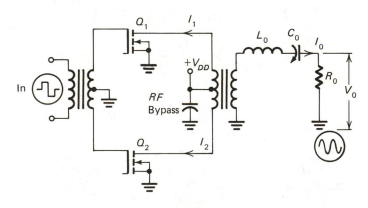

FIGURE 3-11
Basic Arrangement of Class D RF Power Amplifier.
High efficiency is achieved because the power FET's are either on or off, never in an in-between (dissipative) state.

RF amplifiers because the load only utilizes the first harmonic, or fundamental frequency, of the square wave. The higher harmonics are prevented from reaching the load. So the *utilization factor* is not 100%. However, the unused energy is not dissipated as heat in the switching device, and the dc power supply is not called upon to supply such dissipation. The fact remains that the class D amplifier is the most efficient type of all. Figure 3-11 shows the basic class D RF power amplifier.

The question naturally arises as to why the class D mode was virtually never used with electron tubes and has been rarely encountered with bipolar transistors. In the case of tube amplifiers, the departure from an ideal switch robs much of the incentive to operate in the class D mode. This is because of the appreciable plate-cathode voltage drop even when the tube is hard driven. Also, the impedance levels in tube circuits makes it difficult to develop and preserve good square waves at high power levels. Notwithstanding, hams have sometimes inadvertently approached class D operation by overdriving their class B and C amplifiers. Unfortunately, the real benefits of this operational mode accrue only when one makes deliberate efforts to produce very rapid switching transitions.

Bipolar transistors simply have not been sufficiently fast switching elements to produce suitable square waves at repetition rates of 2 MHz or greater. (The very efficient performance obtained from switching-type power supplies generally use switching rates on the order of 20 or 25 kHz.)

It is the advent of the power FET that has made class D operation feasible for frequencies at least up to 30 MHz. These devices, unlike bipolar transistors, are not plagued with minority charge storage problems. Although the drain-source voltage drop exceeds the collector-emitter drop of good RF bipolar transistors, it is still quite low. By using as high a dc voltage as permissible, the effect of the drain-source voltage drop becomes of minor consequence. It is the ability of the power FET to produce nearly instantaneous rise and fall times that renders it a superb device for class D RF service.

There are actually two ways of causing the power FET to develop a square wave in its drain circuit. The most obvious has already been alluded to: a square wave of voltage is applied to the gate, and the drain-source circuit simply follows suit in switching from the on to the off state of conduction. Another technique is also simple, but not quite so obvious. It involves the insertion of a quarter-wave transmission line section in the drain circuit. The gate need not be impressed with a square wave. Rather, half-sine waves can be applied to the gate in much the same manner employed in class B operation. Figure 3-12 illustrates the basic circuitry. Although the power FET is caused to function in an analogous fashion to the class D amplifier, such a circuit is said to operate in the class F mode. By

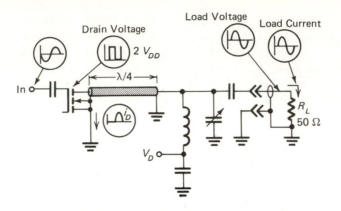

_____ FIGURE 3-12 _____
Basic Arrangement of Class F RF Power Amplifier.
The quarter-wave transmission line causes the drain voltage to be
a square, rather than a sine wave. Because of this, the FET
operates in similar fashion to the class D circuit of Figure 3-11.

referring to the waveforms of Figure 3-13, the operation of the class F amplifier can be explained.

Assume sine-wave input. If it were not for the quarter-wave line, the
drain voltage would be sinusoidal and the FET would operate in class B.
In class B operation, efficiency is limited by, among other things, the fact
that there are times when *both* drain voltage and drain current exist. During such times, there is dissipation in the drain-source circuit of the FET.
If, however, drain voltage could be *converted into a square wave,* drain
voltage and drain current would never coexist. Then their product would
be zero throughout the RF cycle, and there would be no power dissipation
in the drain-source circuit of the FET. Note that essentially the same reasoning applies here as in the class D amplifier of Figure 3-11. It is evident
that the quarter-wave line, although neither an active nor a nonlinear element, exhibits wave-shaping properties.

A quarter-wave line inverts the impedance of its load at odd harmonics of the fundamental frequency. For example, if a quarter-wave line
is loaded by a short-circuit, the generator end of the line will "see" an
open circuit (an infinite impedance) at f, $3f$, $5f$, $7f$, $9f$, and so on. Note that
the fundamental frequency, f, is itself an odd harmonic. (One widely used
manifestation of this transmission-line behavior is the employment of
quarter-wave lines to simulate parallel-resonant circuits in order to provide high impedance at the fundamental frequency.)

Inasmuch as the quarter-wave line is used in *conjunction* with a sim-

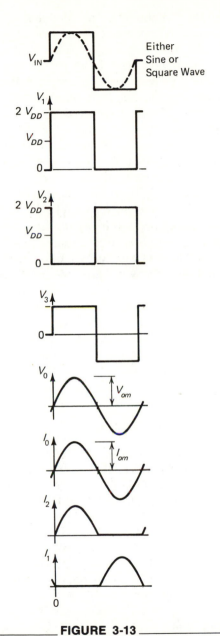

FIGURE 3-13

Wave-Form Diagram of Class D RF Power Amplifier.
Note that the drain voltage of a particular FET is zero while drain
current flows. Such an idealized situation implies zero drain-circuit
dissipation.

ple parallel-tuned tank, it should be recalled that such a parallel-resonant circuit behaves as a high impedance to the fundamental frequency, but as a short-circuit to *all other frequencies*. That is, it supports frequency *f*, but *rejects all its harmonics*. Now let us focus our attention on the drain of the FET and see what happens as a consequence of the *combined* actions of the quarter-wave line and the parallel-resonant *LC* circuit.

All odd harmonics of *f* are supported at the drain, even though they are rejected at the load. This stems from the impedance-inverting property of the quarter-wave line. But *f*, from the load, is short-circuited at the drain. This is because *f* is supported at the load; impedance inversion that makes the load "see" a short at the drain. So we have a situation where all odd harmonics, except *f*, itself, are reflected from the load back to the drain. However, *f* is *already present* at the drain because the quarter-wave line acts as a high impedance at *f* (as seen from the sending end of the line, the drain).

Thus, the drain circuit is caused by the line and *LC* tank to support *f* and its odd harmonics. (At even harmonics of *f*, the line simply reproduces the short-circuit imparted by the *LC* tank.) In this way, the requisite square wave is synthesized at the drain. The synthesis stems from the appropriate *combination* of the Fourier constituents, or "building blocks," of a square wave: a fundamental frequency and a large number of its odd harmonics. (The mathematical justification of this phenomenon also requires fortuitous conditions of magnitude and phase of the harmonics.)

● A Family of MOSFET Giants

The IRF35X family of power MOSFETS made by International Rectifier Corporation are worthy of consideration not merely because of their inordinately high power dissipation ratings (150 W) but because of their simultaneously high dc operating voltages. The 350- and 400-V drain-source voltage rating carried by these devices should greatly ameliorate problems accompanying low-impedance output networks, the scourge of high-power amplifiers using bipolar transistors. Inasmuch as the required impedance of an output network is nearly proportional to the square of the dc operating voltage, these MOSFETS can be implemented with output networks having impedance levels on the order of one hundred and forty times greater than that for a bipolar transistor of similar power capability but operating from a nominal 25-V dc source. Except for lack of the familiar glow of a filament, the nature of such a MOSFET RF amplifier is, indeed, suggestive of tube circuitry.

Admittedly, these power MOSFETS are neither characterized nor intended for RF applications. And it is generally wiser to utilize RF power devices which have been optimized for specific performance parameters.

Electrical Characteristics @ $T_C = 25°C$ (Unless Otherwise Specified)

Parameter		Type	Min.	Typ.	Max.	Units	Conditions
BV_{DSS}	Drain — Source Breakdown Voltage	IRF350 IRF352	400			V	$V_{GS} = 0$
		IRF351 IRF353	350			V	$I_D = 1.0$ mA
$V_{GS(th)}$	Gate Threshold Voltage	ALL	1		3	V	$V_{DS} = V_{GS}, I_D = 1$ mA
I_{GSS}	Gate — Body Leakage	ALL			100	nA	$V_{GS} = 20$V
I_{DSS}	Zero Gate Voltage Drain Current	ALL		0.1	1.0	mA	V_{DS} = Max. Rating, $V_{GS} = 0$
				0.2	4.0	mA	V_{DS} = Max. Rating, $V_{GS} = 0$, $T_J = 125°C$
$I_{D (on)}$	On-State Drain Current	IRF350 IRF351	11			A	$V_{DS} = 25$V, $V_{GS} = 10$V
		IRF352 IRF353	10			A	
$R_{DS (on)}$	Static Drain-Source On State Resistance	IRF350 IRF351		0.25	0.3	Ω	$V_{GS} = 10$V, $I_D = 5.5$A
		IRF352 IRF353		0.3	0.4	Ω	
g_{fs}	Forward Transconductance	ALL	5.0	9.0		S (℧)	$V_{DS} = 100$V, $I_D = 5.5$A
C_{iss}	Input Capacitance	ALL		3000	4000	pF	$V_{GS} = 0$, $V_{DS} = 25$V, f = 1.0 MHz
C_{oss}	Output Capacitance	ALL		400	600	pF	
C_{rss}	Reverse Transfer Capacitance	ALL		100	200	pF	
$t_{d (on)}$	Turn-On Delay Time	ALL		40	60	ns	$I_D = 5.5$A, $E_1 = 0.5$ BV_{DSS}
t_r	Rise Time	ALL		100	150	ns	$T_J = 125°C$ (MOSFET Switching times
$t_{d (off)}$	Turn-Off Delay Time	ALL		300	400	ns	are essentially independent of operating
t_f	Fall Time	ALL		100	150	ns	temperature.)

Thermal Characteristics

$R_{\theta JC}$	Maximum Thermal Resistance Junction-to-Case	ALL		0.83		deg C/W	

FIGURE 3-14

Electrical Characteristics of a Family of High-Power MOSFET Devices.
Although not intended for RF applications, the experimentally inclined will find these devices worthy of consideration.
(Courtesy of International Rectifier Corp.)

However, experimentation is often the precursor of devices and techniques with specific engineering orientations. These MOSFETS should whet the appetite of the experimentally inclined.

The electrical characteristics of this family of power MOSFETS are given in Figure 3-14. Curves depicting the salient operating characteristics are shown in Figure 3-15. One should not be discouraged by the turn-on and turn-off delay times. These apply to switching circuits where time must be alloted to charge and discharge the gate input capacitance. In RF applications, the gate capacitance will be incorporated as part of the input network and there will be no such "delay."

It appears likely that these interesting devices could readily be put to use in narrowband class C amplifiers for service in the high-frequency amateur bands (2 to 30 MHz). Probably, neutralization would be re-

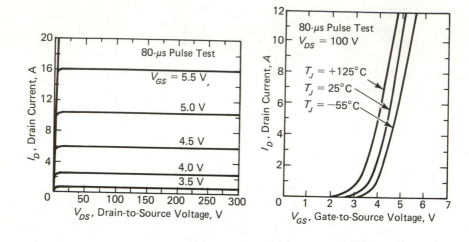

FIGURE 3-15

Basic Characteristics of Giant Power MOSFET's.
Evidenced in these curves are the pentodelike operating mode, the
exceedingly high transconductance, and the possibility of
designing an amplifier circuit with a high-impedance output
network (because of the relatively high voltage operating range
compared to bipolar transistors).
(Courtesy of International Rectifier Corp.)

quired. Also, amateurs have found MOSFETS capable of performing well
in linear amplifiers for SSB transmitters.

The symbology and nomenclature referring to a "junction" should not
be taken to imply a PN region (for example, T_J, $R_{\theta JC}$, and actual use of
the word "junction" in Figure 3-14). This is a carry-over from terminology used in specifying bipolar transistors. For practical purposes, one can
think of internal operating temperature, rather than the alleged "junction" temperature. Another interpretation is to think of the dielectric
"junction" between the gate and the drain-source structure.

● Three Circuit Configurations

Bipolar transistors are useful for RF power applications in the three
circuit configurations, common emitter, common base, and common collector. Correspondingly, tube RF circuits can utilize the common-cathode, common (grounded) grid, and common-plate connections. At the
present writing, the power MOSFET is most frequently seen in RF circuits configured around the common-source connection. A salient feature

of common-source operation is that amplifier stability tends to become better as one goes higher in the VHF and UHF region.

It should not be assumed that other connection modes are ruled out for power MOSFET's. For example, the relative fragility of the gate structure is no impediment to common gate operation if care is exercised in maintaining voltage and current ratings. Although input impedance will then be low, it will be much higher than for analogous operation (common base) of a similarly rated bipolar transistor. However, the stability of common-gate amplifiers tends to be much better in the low- and high-frequency spectrum than at VHF and UHF. Thus, this situation is the opposite of that prevailing with the practical implementation of bipolar transistors. Common-gate amplifiers are noteworthy in that they can exhibit constant and essentially resistive input impedances over many octaves of frequency.

The experimentally inclined should also investigate the attributes of the common-drain circuit. Although this configuration has been largely played down or ignored, a similar situation long endured for RF applications of bipolar power transistors. At least one major semiconductor firm now produces a wide line of bipolar power transistors specifically intended for the common-collector configuration at microwave frequencies. Excellent performance obtains, notwithstanding much literature assigning inferior status to this circuit mode for RF work.

● Disadvantages of the Power MOSFET

Practical experience and basic principles both teach that desirable features in devices are often traded off for some less-than-favorable characteristics. Surely, the impressive behavior of the power MOSFET must be accompanied by *some* disadvantages when compared with the bipolar transistor. Indeed, the following performance shortcomings can be ferreted out:

- The "saturation voltage" is higher than is generally attainable in equivalent RF power bipolars. Thus, saturation voltages in the vicinity of 3 V are not uncommon. This, of course, reduces the effective use of the dc supply and increases device dissipation.

- Input capacitance is relatively high. It is more "visible" than in a bipolar transistor, where the input capacitance tends to be somewhat swamped out by the low resistive component of the input impedance.

- The breakdown mechanism of the gate is damaging. Thus, overdrive, or overbiasing, can puncture the thin oxide-dielectric film of the gate. This contrasts to the junction FET, where gate breakdown is not necessarily injurious. Vulnerability to gate puncture increases with drain voltage.

In addition to gate damage caused by inappropriate operating conditions, the gate is subject to destruction from static electricity during handling. (Once successfully installed in its circuit, the device tends to be relatively immune from the effects of static charges. Care should still be exercised to prevent leakage currents from soldering irons from passing into the gate circuit; vulnerability to damage from this often unsuspected source is much greater than with bipolar transistors.)

Although diode protection of the gate circuit reduces danger at low frequencies, by the time we get into the tens of megahertz region, the side effects and performance degradation caused by protective diodes render them less desirable than a little extra care in handling and operation.

● Avant Garde FET's and RF Techniques

The continuing evolution of power-FET technology is bound to affect high-frequency techniques. Most of the circuit applications of RF power amplifiers using these devices were not extant just a few years ago; one could easily indulge in speculations of the science-fiction kind in prognostications of the near-future. Inasmuch as the theme of this book primarily concerns itself with the practical and the realizable, space will be devoted to neither mere extrapolation of the present state of the art nor to "blue-sky" achievements. However, the following FET devices and techniques are representative of commercially-available devices which have recently (during the production of this book) made their debut. They will be of interest to those motivated to keep pace with new developments and to the experimentally inclined:

- A number of manufacturers have introduced lines of P-channel power-MOSFET's. These devices are analogous to PNP bipolar transistors insofar as concerns the polarity of operating voltages. Surprisingly, these newer devices often have comparable power ratings to established N-channel devices. Even better, they have, in many instances, been designed to have very similar parameters to specified N-channel devices. The availability of such paired-devices makes possible simple implementations to push-pull amplifier circuitry. An example of such a *complementary-symmetry* amplifier is shown in Figure 3-16. Although the FET's connections are more suggestive of a parallel than a push-pull arrangement, actual operation occurs in true push-pull fashion. It is probable that the greatest appeal of such circuitry will be with the designer of ultrasonic equipment. The simple drive requirement and the single-ended output tank configuration also make this application worthy of consideration as a transmitter output amplifier.

● At least the following firms are now marketing P-channel power MOSFET devices: International Rectifier, Siliconix, Supertex, Intersil, ITT, and Westinghouse.

● Overseas companies have developed power MOSFET's which can be driven to full output directly from 5-V TTL logic signals. At the same time, some of these devices have 1000 V drain-source ratings. This suggests relaxations in the design of output networks. More specifically, it appears that final amplifiers utilizing these high-voltage FET's could be associated with pi and pi-L networks designed for tube circuits. This could prove very advantageous in the selection and availability of practically sized inductors and tuning capacitors. Also, it should be relatively easy to obtain class D operation from the square-wave drive signals of TTL logic.

● Another overseas development is a JFET capable of delivering one-kilowatt of ultrasonic power into an appropriate load. It

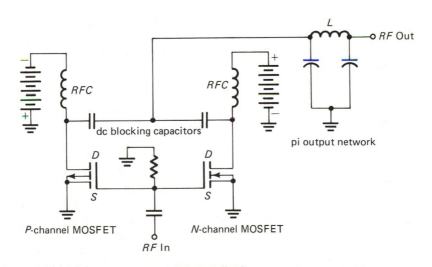

FIGURE 3-16
Complementary Symmetry Push-Pull Amplifier Using MOSFET's.
Although a dual dc supply is needed, this scheme provides benefits of push-pull operation with the simplicity of single-ended drive and output circuitry.

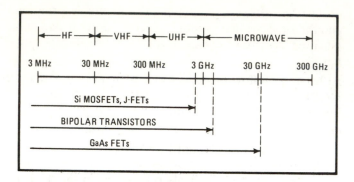

_____ **FIGURE 3-17** _____
The General Frequency Capabilities of Solid-State Power Devices.
Note that the gallium-arsenide FET excels the other devices in the 10 to 30 GHz region of the microwave spectrum.

stands to reason that such a device can also be used for low-frequency, and probably medium frequency, RF power applications. One would expect the frequency capability to at least extent to the AM broadcast band.

• For those whose mission it is to extend RF power performance into ever higher regions of the microwave spectrum, the _gallium-arsenide_ FET is the logical candidate. The frequency capabilities of solid-state power devices is indicated in a general way in Figure 3-17.

In conclusion, it is interesting that FET devices are not only bringing about useful extensions of RF power performance, but are obviously motivating the designers of bipolar power devices to be more innovative than ever before. A certain degree of complacency had set in once bipolar transistors proved their competitive status with certain tubes. Now, because of the unexpected power-frequency capabilities of FET structures, it is clear that bipolar devices must excel in meaningful ways, or atrophy from lack of demand. Yet, it is quite evident that the producers of bipolar transistors do not feel that this venerable device is vulnerable to the technical obsolescence claimed in FET advertisements. What we shall see immediately ahead is enhanced competition between these two solid-state power devices, as well as accelerated improvements in both. Enough time has passed so that it is fairly safe to say there will not be a "winner" _per se_. Rather, each will dominate in various services and applications where unique voltage, current, frequency, cost, and other factors influence designers and hobbyists.

4

Impedance-Matching Networks

When implementing solid-state RF power circuits, it is not commonly found that one can simply "throw in" a resonant circuit and obtain satisfactory results. RF power transistors display very *low* input and output impedances compared to tubes. Both the resistive and reactive components of these impedances vary considerably with frequency and with operating conditions. It turns out that RF oscillators and amplifiers can be designed and constructed to provide reasonably acceptable performance only if two mutual requirements are met: (1) one must have an appropriate solid-state device, and (2) one must interface it with appropriate impedance-matching networks. There are other stumbling blocks, to be sure. But these two are paramount. Although the topic of impedance-matching networks can easily fill the pages of a voluminous text, the material covered in this chapter should suffice for many of the everyday situations encountered in solid-state RF power.

● Basic Considerations

Some aspects of impedance matching have already been alluded to in Chapter 2. This is because the operation of the solid-state RF generator or amplifier is very dependent upon its associated LC tank circuits. In this chapter we will examine various ways in which optimum performance can be achieved via the selection of appropriate input and output circuits. Generally, "optimum performance" will simply be the most reasonable compromise of several contradictory parameters. For example, high Q is always desirable from the standpoint of harmonic rejection. Unfortunately, overall efficiency suffers as Q is raised. We refer here to the operating Q, that is the "loaded" Q of the LC circuits. (The Q of *individual* inductors and capacitors in a network should always be as high as is practically and economically feasible.) The reason that efficiency decreases as the operating Q is made higher is due to the increased losses produced by the higher circulating currents between the L and C elements.

Because of efficiency considerations, the operating Q of the output circuit is often designed to be about 10 or 12. The simple fact of life here is that reasonable frequency selectivity is provided by such a tuned circuit without seriously degrading the efficiency of the amplifier. This may pose another conflict, however. The use of practically sized L and C elements may not always be within ready reach when we peg the operating Q in the vicinity of 10 or 12. One consequence of this is that transistor amplifiers operating at powers much above the flea level cannot use the familiar pi output circuits which have worked so very effectively with our tube amplifiers.

On the other hand, input networks to transistors may have lower Q's, say in the 2 to 6 range. Even lower Q's are used when broadbanding is desirable. Low-Q input networks contribute to stability and makes tune-up and adjustment less critical. A high-Q input circuit will not contribute much to wave purity and harmonic rejection in the output circuit of class B and C amplifiers. The main function of the input circuit is impedance matching. In this regard, it is necessary to match the requirements of both the driver and the driven stage. The input circuit impedance of a transistor is inherently low and invariably contains an appreciable reactive component. The reactive component tends to be capacitative at low and medium frequencies, but is often inductive at VHF and UHF. However, at these higher frequencies, the manufacturer often utilizes the "internal" reactances to provide a "built-in" impedance-matching network. Nonetheless, external network elements, such as transformers, may still be required.

● General Design Approach

The rigorous calculation of impedance-matching networks can lead to considerable complexity and cumbersome mathematics. In practical RF power work, such involvement is, fortunately, not usually needed. We do have to get within reasonable limits. Thereafter, one can readily optimize operation by experimenting or by providing variable or tapped elements. Such a combination of analytical and empirical approaches is justifiable if for no other reason that we seldom find ourselves in a position to know the values of all the necessary parameters. For example, the antenna feedline may be assumed to be 50 ohms. Such an assumption is just that; in most cases there will be appreciable departure from the ideal resistive 50 ohms. Generally, the accompanying reactive component must be somehow tuned out. This is often a task for the output network. Thus, variable elements would be necessary even if great pains were taken to design a rigorously accurate network. This is especially true if, as is often the case, more than a single fixed frequency is involved.

There are yet other reasons why simplified design approaches are desirable. The tolerances on transistor parameters are notoriously sloppy. This, alone, would generally necessitate some departure from rigorously designed networks. And, of course, the capacitances and inductances of the network elements generally involve tolerances on the order of $\pm 10\%$. Nor should we forget the influence of stray parameters, especially at higher frequencies. And any deviation from a set operating voltage and current causes changes in the input and output impedances of a transistor. All things considered, our objective will be to keep things simple. In pursuit of this philosophy, all output networks will be considered to have an operating Q of 12, and all output loads will be considered to be 50 ohms and to be purely resistive. Moreover, mathematical shortcuts will be resorted to whenever the ultimate result is not drastically affected. In general, one or more provisions for tuning or adjusting the individual network elements will be required. Perhaps, in some cases, the elements can ultimately be "frozen" at fixed values after appropriate experimentation. Instead of merely citing the relevant equations, a practical example of each technique will be worked out.

● Basic Mechanism of Impedance Transformation

The manner in which a simple LC circuit can act as an impedance transformer is not obvious from an initial inspection. This exceedingly useful behavior actually stems from a unique relationship between series and parallel resistance-reactance circuits. It happens that every parallel XR

circuit can be duplicated by an *equivalent* series circuit. What is implied by such equivalency, and what is its significance with regard to the function of impedance transformation?

Shown in Figure 4-1 is a parallel *LR* circuit together with a possibly equivalent series *LR* circuit. Equivalency is said to exist between these circuits when the impedances seen across their respective terminals are the same. In the general case, this condition obtains at a certain frequency and in such a way that R_p differs from R_s. For practical purposes, X_p and X_s can be considered as equal. The notion of equivalency derives from the fact that were these two circuits to be put in a "black box" the magnitude and phase angle of their impedances would be the same.

What has been stated about the *LR* circuits of Figure 4-1 applies also to the *CR* circuits of Figure 4-2. Here, again, a frequency can be found in which equivalency exists. (Or, conversely, at a given frequency, the elements can be manipulated to produce equivalency.) In both the inductive

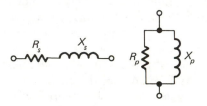

FIGURE 4-1
**Series and Parallel Circuits Which Can Be Made
to Have Similar Behavior.**
By appropriately selecting the element values, both circuits can be
caused to display the same impedance magnitude and phase
angle.

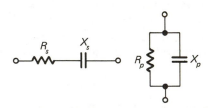

FIGURE 4-2
Circuits Which Like Those of Figure 4-1 Can Exhibit Equivalency.
Here, also, by appropriate selection of element values the two
circuits can be caused to have the same impedance level and
phase angle.

and capacitive circuits, the equations defining equivalency are as follows:

$$R_p = R_s(Q^2 + 1)$$

$$X_p = X_s \frac{(Q^2 + 1)}{Q^2}$$

But in many practical situations where Q is 5 or higher, we can use the following simplifications:

$$R_p = Q^2 R_s$$
$$X_p = X_s$$

Also relevant are the equations for Q and for the specific types of reactance:

$$Q \text{ for the series circuit } = \frac{X_s}{R_s}$$

$$Q \text{ for the parallel circuit } = \frac{R_p}{X_p}$$

$$\text{Inductive reactance: } X_L = 2\pi f L$$

$$\text{Capacitive reactance: } X_C = \frac{1}{2\pi f C}$$

Not only can equivalencies be devised from series and parallel circuits containing one kind of reactance; they can also be devised from such circuits containing opposite types of reactance, as in Figure 4-3. As be-

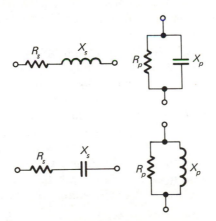

FIGURE 4-3

Equivalent Circuits Using Opposite Reactances.
In each of these two cases, the series circuit can be made to have the same impedance magnitude and numerical phase angle as the corresponding parallel circuit.

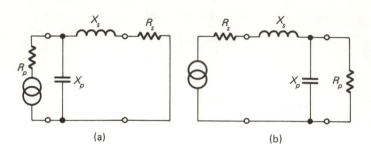

(a)   (b)

_____ **FIGURE 4-4** _____
Two Versions of the Basic _L_ Network.
(a) Network for transforming high resistance (impedance), R_p, to a
lower value, R_s. (b) Network for transforming low resistance
(impedance), R_s, to a higher value, R_p. In both networks,
$$X_s/R_s = R_p/X_p$$

fore, the series resistance R_s tends to be lower than the parallel resistance,
R_p, when equivalency between the two circuits prevails. Having gone this
far, one may well ponder the consequences of connecting the series and
parallel network together in order to form a single network. Providing this
can be done, it should be apparent that such a network would have the
ability of transforming a low resistance, such as R_s, to a higher resistance,
R_p (or, conversely, a high resistance, R_p, to a low resistance, R_s). It should
now be obvious that we have "evolved" the _L_ network. As will shortly be
seen, _L_ networks possess other features besides the mere ability to transform between resistive levels.

It should be pointed out that these concepts of _L_ network operation
are not likely to be found in texts dealing with filters. The _L_ section is,
indeed, a true low-pass filter when configured with series and shunt arms
as shown in Figure 4-4. However, the emphasis is on a different operating
mode when the network is employed primarily for its filtering action. In
such instance, the operating frequency is below the resonant (cutoff) frequency, and the input and output resistances are both equal to the "characteristic resistance" of the filter. An example of such operation is a TVI
filter inserted at the output of a transmitter. Whereas such filters are often
pi networks, the pi network is, itself, merely a cascade of two or more
basic _L_ sections.

Conversely, the operating mode of the _L_ network when used for
transformation between unlike resistance or impedance values requires
the condition of resonance between its inductive and capacitive reactances. The price paid for the ability to match unlike input and output resistances in such transformerlike fashion is that the network then be-

comes a single-frequency device. In practice, a band of frequencies can be accommodated because the operating Q is not high, as selective circuits go. The notion of viewing L network as *combined* series and parallel RX circuits provides insights not readily attained from filter theory.

The interesting and significant aspect of equivalency is that the Q's of the parallel circuit and the series circuit are equal. This leads to the important equality, $X_s/R_s = R_p/X_p$. This relationship enables calculation of the element values. Only Q and one of these four parameters need be known to calculate the remaining two.

Now let us again look at the L network such as shown in Figure 4-4(a). This is actually an LCR circuit, but the concepts of equivalency described for the simpler circuits apply here, too. What we have is a resonant tank which looks like a parallel-resonant circuit at the input and a series-resonant circuit at the output. Again, we postulate that the series-resonant Q and the parallel resonant Q have the same value. Mathematically, this network will "work" with a high value of R_p and a low value of R_s. Another condition required is that $X_p = X_s$, which is tantamount to resonance. Thus, at resonance the concept of equivalency is satisfied in one physical circuit, and a high resistance R_p is transformed down to a low resistance R_s.

Such an L network is even better than the foregoing logic indicates, for it is not merely a transformer, but *also* a low-pass filter. Moreover, it can be used in inverse fashion, as shown in Figure 4-4(b), for transforming a low value of R_s to a high value of R_p. Yet another aspect of the L network is that it can "accommodate" reactance present in R_p and in R_s. It does this by absorbing such reactance in its own series and shunt arms. In so doing, the resonant condition occurs with some departure from calculated element values, but the desired transformation takes place. (In practice, the amount of reactance that can be absorbed in this manner must not be stretched too far, for then the readjustment to resonance can cause an appreciable change in operating Q, or, in other instances, resonance will not be readily attainable.)

Last but not least, the L network is the basic "building block" of pi and tee networks. As with the L network, the pi and tee networks accomplish impedance transformation by simultaneously acting as parallel- and series-resonant tanks. And they, too, are low-pass filters. As such, their harmonic attentuation tends to be more effective than the simple two-element L network.

● Derivation of Basic Pi Network

Figure 4-5 shows a pi network drawn in such a manner that it can be readily seen to be comprised of two cascaded L networks of the configurations illustrated in Figure 4-4(a) and (b). Resistances R_{s1} and R_{s2} are

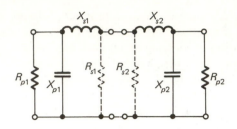

_____ FIGURE 4-5 _____
**Schematic Diagram Depicting Pi Network
As Two Cascaded L Networks.**
Although the parallel resistances, R_{s1} and R_{s2}, are physically
nonexistent, they have mathematical significance during the
design procedure.

shown in dashed lines because they are physically nonexistent. These resistances are used during the calculation of the pi network because each L
section is dealt with individually before the two are "joined." The completed pi network operates as if a resistance, R_{ss}, were connected across
the mid-section of the pi network. R_{ss}, of course, is the value of the parallel combination, R_{s1} and R_{s2}. That is,

$$R_{ss} = \frac{R_{s1} \times R_{s2}}{R_{s1} + R_{s2}}$$

Appropriately, R_{ss} is known as a *virtual* resistance.

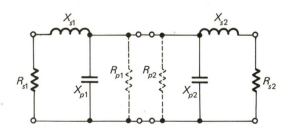

_____ FIGURE 4-6 _____
**Schematic Diagram Depicting Tee Network
As Two Cascaded L Networks.**
Although the parallel resistances, R_{p1} and R_{p2}, are physically
nonexistent, they have mathematical significance during the
design procedure.

Another consequence of the joining of the individually calculated L sections is that X_{s1} and X_{s2} combine to produce a single physical inductor. It is well to note here that inductive reactance and capacitive reactance are used in the element designations. These reactances are ultimately converted to inductance and capacitance. This approach tends to systematize design procedure. (The design procedure will be illustrated in a forthcoming example.)

● Derivation of Basic Tee Network

Figure 4-6 shows a tee network drawn in such a manner that it also can be readily seen to be comprised of two cascaded L networks of the configuration illustrated in Figure 4-4(a) and (b). Resistances R_{p1} and R_{p2} are shown in dashed lines because they are physically nonexistent. These resistances are used during the calculation of the tee network because each L section is dealt with individually before the two are "joined." The completed tee network operates as if a resistance, R_{pp}, were connected across the mid-section of the tee network. R_{pp}, of course, is the value of the parallel combination, R_{p1} and R_{p2}. That is,

$$R_{pp} = \frac{R_{p1} \times R_{p2}}{R_{p1} + R_{p2}}$$

Appropriately, R_{pp} is known as a *virtual* resistance.

Another consequence of the joining of the individually calculated L sections is that X_{p1} and X_{p2} combine to produce a single physical capacitor. The most straightforward design of the tee network employs inductors in the series arm which are not coupled to one another. (It is feasible to use a single tapped inductor, but it is then necessary to bring the coupling coefficient into the calculations. Often it is then more practical to employ empirical techniques.) The design procedure of the tee network will be illustrated in a forthcoming example.

● Wave Purity in Transistor Amplifiers

It is not as easy to obtain acceptable wave purity from a transistor amplifier as it is from its tube counterpart. Of the two devices, the transistor takes first prize for imperfect switching operation, nonlinear dynamic performance, parametric (varactor) behavior, and anomalies involving negative resistance, discontinuities, and the like. As if this were not enough, it often displays the tube's vulnerability to UHF and VHF parasitic oscillation, together with its own unique tendency to generate spurious low-frequency oscillation. This follows from the fact that most RF power transistors are operated far beyond beta cutoff frequency. This implies an increase of 6 decibels (dB) in gain for each halving of frequency. It is not uncommon for the rising low-frequency gain to provoke spurious oscillations in the audio-frequency range. Most of these amplifier circuits utilize the common-emitter configuration. It might be supposed that the low-frequency troubles could be circumvented by using the common-base circuit. However, one then becomes involved in an undesirable trade-off: the already low input impedance of transistors is the lowest for the common-base arrangement, and new difficulties are then experienced with respect to driver and input network design. The common-base *oscillator* is, however, useful at UHF and the microwave region.

Yet another source of frequency contamination in the output of transistor amplifiers is feedthrough from frequency-multiplier stages. This is aggravated by the fact that transistors are less effective as isolating elements than are tubes. (This is especially so in frequency multipliers, which, as is commonly the case, dispense with neutralizing circuits. Neutralization is generally unnecessary because of the relatively low power gain developed in the beta cutoff region and because the input and output circuits are tuned to different frequencies.) The low gain necessitates high drive power, which, unfortunately, worsens the feedthrough of the sub-harmonic frequencies.

Other things being equal, the use of push-pull output stages reduces second and other even-order harmonics and thereby relaxes the filtering burden of the output network. Simply making the output network more complex often leads to awkward situations where different frequencies or antennas must be accommodated. A common practice is to use a simple output network which functions primarily to bring about an effective impedance transformation. Then, in the interest of wave purity, a low-pass or band-pass filter is inserted in the antenna feeder line.

● Selection of Network Types for Transistor RF Amplifiers

The previous discussions of impedance-matching networks have been in the nature of an academic prelude for the purpose of developing basic insights. We may now present practical examples of network designs. From these, the experimenter or designer can readily adapt the procedures to specific situations by merely substituting the relevant numbers. As previously stated, the emphasis will be on approximate solutions via simplified equations. Experience teaches that this approach generally produces near-optimum results. Whatever modification is thereafter deemed desirable will, in most cases, involve no greater departures from first trial design than would pertain to more rigorous computations.

Besides the basic L, pi, and tee networks, there are many modifications of these in which certain performance parameters, such as broadbanding, tuning convenience, harmonic attenuation, and so on, are optimized. In general, the L network provides impedance transformations from high to low values, and vice versa. The tee network is particularly well suited to transforming between two relatively low impedances. The pi network is probably best adapted to transform between relatively high impedances, but as has been evident from tube amplifiers, it can also serve well in transforming between a high and low impedance. What generally decides these matters in practice is the size, practicability, availability, and cost of the network elements. These vary greatly with application, frequency, and power level. While it may be possible to design more than one network to comply with the requirements of a particular circuit application, the actual implementation will usually favor one over the other. This is especially true when one must also consider the voltage and current capabilities of available or readily constructed inductors and capacitors.

Except where very low power levels are involved, the requirement for impedance transformation at the output of the transistor is opposite that of tubes; that is, it is necessary to transform from a *lower* to the higher 50-ohm impedance of the antenna feedline. Another aspect of transistor amplifiers is the inordinately low input impedance presented by the base-emitter circuit. (Somewhat excepted from this statement is the power FET, with its tubelike input impedance.) Also bearing on network design is the high input capacitance of bipolar transistors throughout the 2- to 30-MHz range and the fact that the collector output capacitance varies with the applied voltage. The latter feature enhances harmonic production. This may be good for frequency multipliers, but may add burden to the filtering task of the output network in "straight-through" amplifiers.

● **Some Facts Pertaining to All Networks**

Although various assumptions and approximations lead to success in most practical design approaches, there is one assumption that generally should not be made. It should not be blandly supposed that element values measured at one frequency yield the same results at some *other* frequency. The more widely displaced the measured frequency and the frequency involved in application, the greater is the probability of malperformance of the network. Also, the higher the operating frequency, the greater is the likelihood that trouble will be encountered owing to element values being other than what had been measured. Lead inductance of capacitors is one of the major contributors to this phenomenon. It is well known that every capacitor has its own resonant frequency and that at higher frequencies the capacitor begins to look like an inductor. This is readily demonstrated with a grid-dip meter and a random selection of ceramic, mica, or other capacitors. If the leads are shorted and the loop thereby formed is coupled to the grid-dip meter, resonances throughout a wide frequency range can be detected. This infers that capacitors begin to "lose" capacitance as resonance is approached from the low-frequency end of the spectrum.

In addition to the simple effect of lead inductance, more complex interplays of stray parameters can assert themselves at frequencies other than the measured frequency. This is true for inductors as well as capacitors. Indeed, inductors often exhibit alternate series and parallel resonances at frequencies which may or may not be simply related to one another. At VHF and UHF, the strays associated with adjacent surfaces often make it mandatory that the *network* be "proved" in its actual operating environment. In any event, L and C values should be directly measured at the intended operating frequency.

The output network of a transistor power amplifier often will provide inadequate filter action for harmonics. It should not be assumed that a high-Q antenna will then suffice for this task. The constants of an antenna are distributed rather than "lumped." This permits the antenna to be responsive to other than its "resonant" frequency.

Inductors using ferrite, powdered iron, or molybdenum cores will have different inductances at different frequencies because the magnetic permeability of such substances is not constant with respect to frequency. Also, the effective inductance of such inductors can change considerably if they are carrying direct current, or if they are operated in the region approaching magnetic saturation.

● Output Impedance of Bipolar Transistors

As with tube amplifiers, we must know how the transistor appears to the output tank or network. To a first approximation, the transistor collector circuit displays an impedance which is determined in the same way as with tubes. That is, the approximate output impedance of a power transistor is simply $V^2/2P$, where V is the applied dc voltage and P is the anticipated power output. This approximation is valid for class C amplifiers.

This expression is dictated by Ohm's law considerations and does not derive from the intrinsic characteristics of the transistor itself. It tends to give values of output resistance which err on the high side. There are two reasons for this. First, the peak values of the RF wave cannot actually be twice the applied dc voltage. The collector saturation voltage prevents this. Collector saturation voltage, V_{CE}, is often in the vicinity 1.5 V or so. Therefore, when the power supply provides 12 V, the maximum collector dc voltage is 10.5 V. Also, the relatively low Q output networks do not quite produce the magnitude of RF peaks which are assumed under ideal resonant circuit behavior. All things considered, it might be well to express the output impedance of the transistor as $\frac{V^2}{2P}$ for the class C amplifier, or as $\frac{V^2}{1.57P}$ for class B amplifiers. For class A amplifiers, $\frac{V^2}{1.3P}$ yields better results as approximation to output impedance.

The output transistor also has a capacitive component. It is not easy to determine this from all data sheets. For one thing, small-signal specifications or static measurements do not suffice for the dynamic conditions encountered in RF power amplifiers. Actually, the output capacitance varies in varactor fashion during the excursion of the RF cycle. Therefore, the best one can do is to postulate an average value. Values within the range of 20 to 60 pF are often found. At VHF and UHF, it is often necessary to explore the effect of output capacitance in a breadboard model closely simulating the physical layout of the final design.

Those with tube experience in RF power amplifiers can acquire the feel of solid-state versions by realizing that the resistive portion of the transistor's output impedance is what a hypothetical electron tube would exhibit if it could operate at several tens of volts and at plate currents measured in amperes.

● Basic Impedance-Matching Networks Used with Transistor Amplifiers

The five LC networks illustrated in Figure 4-7 represent the majority of lumped circuit impedance-matching networks used with transistor amplifiers operating at power levels above several watts. (At lower power levels, parallel-tuned tanks with tapped or secondary windings are often encountered.) In these networks, one of the impedances is considered to be 50 ohms resistive. The other impedance is considered to have a resistive and a reactive component. Although the reactive component is depicted as capacitive, inductive reactance is also encountered, for example at the input (base) of VHF and UHF transistors. The practical implication is that R_1 can be considered to be the output impedance of the transistor, whereupon R_L will be the idealized antenna feedline impedance. Conversely, R_1 may, in some instances, be considered to be the input impedance of the transistor, whereupon R_L will be the driving source impedance, that is, the output impedance of the driver amplifier. For sake of simplicity, let us first consider R_1 to be the resistive component of the output impedance of the transistor.

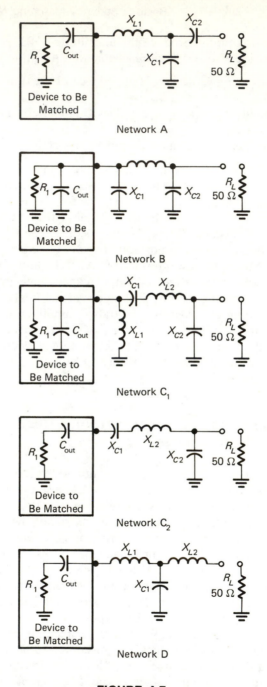

FIGURE 4-7

Basic Impedance-Matching Networks Used
With Transistor RF Amplifiers.

Convenience of computation dictates whether the device to be
matched is treated as a series or a parallel circuit.

It will be observed that R_1 is always accompanied by capacitance, C_{out}, implying that the commonly used formula for output resistance,

$$R_{out} = \frac{V^2}{2P}$$

is an approximation only. Note also that R_1 and C_1 are shown as a parallel circuit in some networks and as a series circuit in others. This is not an inconsistency, but results from the fact that the parallel circuit serves the purpose of computational convenience in some cases, whereas the series circuit facilitates computation in other network situations. The thing to keep in mind is that the series and parallel circuits are convertible by the equations previously given, which are repeated in Table 4-1. Indeed, in transistor specifications, the input and output impedances of RF transistors may be given or displayed in several other ways. This may appear confusing. However, the various notational schemes all represent the same information, and any one of them can be converted to any of the others. More specifically, all can be converted into equivalent parallel or series RC circuits so that R_1 and C_{out} in the networks in Figure 4-7 can be assigned definite values.

The reason for the additional impedance formats is that they provide advantages during the actual laboratory measurement procedure. They will be discussed in subsequent paragraphs.

● Formulas for the Basic Impedance-Matching Networks

The "crank-grinding" formulas for solving the networks illustrated in Figure 4-7 are given in Table 4-1. These formulas are predicated upon the following procedures:

- Q is selected from other considerations. Such selection is predicated on bandwidth, harmonic rejection, and operating efficiency of the amplifier. In the real world, another governing factor enters into the selection of Q—the *practicability* of elements comprising the network. Inductors of a few nanohenries or of several hundred millihenries might prove difficult or costly to implement. Similarly, capacitors of a few picofarads or several thousand picofarads might involve practical difficulties not revealed by the network formulas.

- R_1 is known and is usually taken as 50 ohms.

- The formulas deal with X_C, rather than C, and X_L, rather than L. This produces a generalized solution. To obtain C and L for the specific frequency (or mid-frequency) involved, use the relationships, $C = \dfrac{1}{2\pi f(X_C)}$ and $L = \dfrac{X_L}{2\pi f}$.

TABLE 4-1

Formulas Applicable to the Solution of the Basic Impedance-Matching Networks

---

When Q is 5 or greater, considerable simplification results from using Q^2 rather than $(Q^2 + 1)$ wherever this term appears. Better still, much arithmetical work can be saved by resorting to the computerized solutions listed in Table 4-2 following.

To convert a parallel resistance and reactance combination to series:

$$R_s = \frac{R_P}{1 + (R_P/X_P)^2}$$

$$X_s = R_s \frac{R_P}{X_P}$$

To convert a series resistance and reactance combination to parallel:

$$R_P = R_s[1 + (X_s/R_s)^2]$$

$$X_P = \frac{R_P}{X_s/R_s}$$

To solve network A (Figure 4-7), select a Q:

$$X_{L1} = QR_1 + X_{C\,out}$$

$$X_{C2} = AR_L$$

$$X_{C1} = \frac{(B/A)(B/Q)}{(B/A) - (B/Q)} = \frac{B}{Q - A}$$

where $A = \sqrt{\left[\dfrac{R_1(1 + Q^2)}{R_L}\right] - 1}$

$$B = R_1(1 + Q^2)$$

To solve network B, select a Q:

$$X_{C1} = R_1/Q$$

$$X_{C2} = R_L \sqrt{\frac{R_1/R_L}{(Q^2 + 1) - (R_1/R_L)}}$$

$$X_L = \frac{QR_1 + (R_1R_L/X_{C2})}{Q^2 + 1}$$

To solve network C_1, select a Q:

$$X_{L1} = X_{C\,out}$$

$$X_{C1} = QR_1$$

$$X_{C2} = R_L \sqrt{\frac{R_1}{R_L - R_1}}$$

$$X_{L2} = X_{C1} + \left(\frac{R_1R_L}{X_{C2}}\right)$$

To solve network C_2, select a Q. L_1 is not used in this network:

$$X_{C1} = QR_1$$

$$X_{C2} = R_L \sqrt{\frac{R_1}{R_L - R_1}}$$

$$X_{L2} = X_{C1} + \left(\frac{R_1R_L}{X_{C2}}\right) + X_{C\,out}$$

To solve network D, select a Q:

$$X_{L1} = (R_1Q) + X_{C\,out}$$

$$X_{L2} = R_LB$$

$$X_{C1} = \frac{(A/Q)(A/B)}{(A/Q) + (A/B)} = \frac{A}{Q + B}$$

where $A = R_1(1 + Q^2)$

$$B = \sqrt{\left(\frac{A}{R_L}\right) - 1}$$

- Networks A, B, and D involve the term $(Q^2 + 1)$ or $(1 + Q^2)$. Where the selected Q is 5 or greater, this term can be simplified to Q^2 in the computations. This will alleviate some of the tediousness in performing these calculations.

● Computerized Network Solutions

The formulas listed in Table 4-1 for the determination of element values in the basic impedance-matching networks involve straightforward algebraic manipulations, but are admittedly tedious to use. Inasmuch as we will probably resort to slide rules, calculators, or tables, an even more elegant approach would be to program a digital computer to produce solutions encompassing many Q selections for all these networks. Such computerized solutions are given in Table 4-2. Here, again, we select Q according to our emphasis on broadbanding, harmonic rejection, operating efficiency of the amplifier, and practicality of the elements. Some intuition pays dividends here; even the computer cannot comment on the practicality of its solutions.

These computerized solutions stem from the formulas listed in Table 4-1. Therefore, the solutions are general in that they apply for all frequencies. To arrive at a specific solution, the inductive and capacitive reactances must be converted to actual L and C values. This is done via the relationships $L = X_L/2\pi f$ and $C = 1/2\pi f X_C$, where f is the frequency of interest. Where considerable broadbanding exists because of low Q, f is the mid-frequency and is actually the geometric mean between the band-edge frequencies, f_1 and f_2. That is $f = \sqrt{f_1 \times f_2}$. The bandwidth is given by f/Q and is the frequency spread between f_1 and f_2. (Band edges f_1 and f_2 are actually 3 dB down, but we need not concern ourselves with this because this condition automatically ensues from the preceding definitions.)

For a given network, certain interpolations can be made to increase the usefulness of the computerized solutions. For example, if it is desired to design network D for a Q of 15, the X_L values would have reactances of 15/10 or 1.5 times the reactances indicated for $Q = 10$. Extrapolation requires caution, and it is best to experiment first with actual indicated values before assuming that certain proportional relationships exist.

● Transistor Impedance Information

To design impedance-matching networks, it is necessary to have at hand the output and/or the input impedance of the transistor under operating conditions reasonably close to those intended. This information is presented in transistor specifications sheets in several different ways, and it would be quite natural to initially experience confusion. However, the different data presentations are not the result of the manufacturer's whims, even though 100% consistency is not attained. There is more than one way to indicate impedance values. But, generally, the reasons for depicting the information in a certain format do not stem from arbitrary decisions. At different frequency ranges, and under different operating conditions, the manufacturer finds it is more straightforward to employ one system of measurement rather than others. This is because of the very

TABLE 4-2

Computerized Network Solutions[a]

NETWORK A

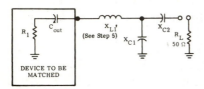

DEVICE TO BE
MATCHED

TO DESIGN A NETWORK USING THE TABLES

1. Transform the parallel impedance of the device to be matched to series form ($R_1 + jX_{C_{out}}$).
2. Define Q, in column one, as X_{L1}/R_1.
3. Choose a Q.
4. For a Q, find the R_s to be matched in the R column and read the reactive value of the components.
5. X_{L1}' is equal to the quantity X_{L1} obtained from the tables plus $|X_{C_{out}}|$.
6. This completes the network.

Q	X_{L1}	X_{C1}	X_{C2}	R_1
1	26	65	10	26
1	27	75.3	14.14	27
1	28	85.68	17.32	28
1	29	96.66	20	29
1	30	108.5	22.36	30
1	32	136	26.46	32
1	34	170	30	34
1	36	213.8	33.16	36
1	38	272.5	36.05	38
1	40	355	38.7	40
1	42	479	41.23	42
1	44	686.32	43.59	44
1	46	1102	45.83	46
1	48	2351	48	48
2	22	32.7	15.8	11
2	24	38.6	22.4	12
2	26	45	27.4	13
2	28	51.2	31.6	14
2	30	58	35.4	15
2	32	65.3	38.7	16
2	34	73.1	41.8	17
2	36	81.4	44.7	18
2	38	90.3	47.4	19
2	40	100	50	20
2	42	110.4	52.4	21
2	44	122	55	22
2	46	134	57	23
2	48	147	59	24
2	50	161	61	25
2	52	177	63	26
2	54	194	65	27
2	56	213	67	28
2	58	233	69	29
2	60	256	71	30
2	64	310	74	32
2	68	377	77	34
2	72	464	81	36
2	76	582	84	38
2	80	746	87	40
2	84	995	89	42
2	88	1409	92	44
2	92	2241	95	46
2	96	4739	97	48
3	18	23.5	22.3	6
3	21	29.6	31.6	7
3	24	35.9	38.7	8
3	27	42.7	44.7	9
3	30	50	50	10
3	33	57.8	54.8	11
3	36	66	59	12
3	39	75	63.2	13

Q	X_{L1}	X_{C1}	X_{C2}	R_1
3	42	84	67	14
3	45	95	71	15
3	48	105	74	16
3	51	117	77	17
3	54	130	81	18
3	57	143	84	19
3	60	158	87	20
3	63	173	89	21
3	66	190	92	22
3	69	209	95	23
3	72	228	97	24
3	75	250	100	25
3	78	274	102	26
3	81	299	105	27
3	84	327	107	28
3	87	358	110	29
3	90	393	112	30
3	96	473	116	32
3	102	575	120	34
3	108	706	124	36
3	114	882	128	38
3	120	1129	132	40
3	126	1502	136	42
3	132	2124	140	44
3	138	3372	143	46
3	144	7119	146	48
4	12	13.2	7.1	3
4	16	20	30	4
4	20	26.9	41.8	5
4	24	34.2	51	6
4	28	42.1	58.7	7
4	32	50.6	66	8
4	36	60	72	9
4	40	69	77	10
4	44	80	83	11
4	48	91	88	12
4	52	103	92	13
4	56	115	97	14
4	60	129	101	15
4	64	144	105	16
4	68	159	109	17
4	72	176	113	18
4	76	194	117	19
4	80	214	120	20
4	84	235	124	21
4	88	257	127	22
4	92	282	131	23
4	96	308	134	24
4	100	337	137	25
4	104	368	140	26
4	108	403	143	27

Q	X_{L1}	X_{C1}	X_{C2}	R_1
4	112	440	146	28
4	116	482	149	29
4	120	527	152	30
4	128	635	157	32
4	136	770	162	34
4	144	945	168	36
4	152	1180	173	38
4	160	1510	177	40
4	168	2007	182	42
4	176	2837	187	44
4	184	4500	191	46
4	192	9497	196	48
5	10	10.8	10	2
5	15	18.3	37.4	3
5	20	26.3	52	4
5	25	34.8	63.2	5
5	30	44	73	6
5	35	54	81	7
5	40	65	89	8
5	45	76	96	9
5	50	88	102	10
5	55	101	108	11
5	60	115	114	12
5	65	130	120	13
5	70	146	125	14
5	75	163	130	15
5	80	181	135	16
5	85	201	140	17
5	90	222	145	18
5	95	245	149	19
5	100	269	153	20
5	105	295	157	21
5	110	323	162	22
5	115	354	166	23
5	120	387	169	24
5	125	423	173	25
5	130	462	177	26
5	135	505	181	27
5	140	553	184	28
5	145	604	188	29
5	150	662	191	30
5	160	796	198	32
5	170	965	204	34
5	180	1184	210	36
5	190	1477	217	38
5	200	1890	222	40
5	210	2510	228	42
5	220	3548	234	44
5	230	5628	239	46
5	240	11874	245	48

([a] *Courtesy of Motorola Semiconductor Products, Inc.*)

TABLE 4-2 (Cont'd.)

Q	X_{L1}	X_{C1}	X_{C2}	R_1
6	12	13.9	34.6	2
6	18	22.7	55.2	3
6	24	32.2	70	4
6	30	42.5	82	5
6	36	53.6	93	6
6	42	65.5	102	7
6	48	78	110	8
6	54	92	119	9
6	60	107	126	10
6	66	122	133	11
6	72	139	140	12
6	78	157	147	13
6	84	176	153	14
6	90	197	159	15
6	96	219	165	16
6	102	242	170	17
6	108	267	175	18
6	114	295	181	19
6	120	324	186	20
6	126	355	191	21
6	132	389	195	22
6	138	426	200	23
6	144	466	205	24
6	150	509	209	25
6	156	556	214	26
6	162	608	218	27
6	168	664	222	28
6	174	727	226	29
6	180	795	230	30
6	192	957	238	32
6	204	1160	246	34
6	216	1422	253	36
6	228	1775	260	38
6	240	2270	267	40
6	252	3015	274	42
6	264	4260	281	44
6	276	6755	287	46
6	288	14250	294	48
7	14	16.7	50	2
7	21	26.8	71	3
7	28	38	87	4
7	35	50	100	5
7	42	63	112	6
7	49	77	122	7
7	56	92	132	8
7	63	108	141	9
7	70	125	150	10
7	77	143	158	11
7	84	163	166	12
7	91	184	173	13
7	98	206	180	14
7	105	230	187	15
7	112	256	193	16
7	119	283	200	17
7	126	313	206	18
7	133	344	212	19
7	140	379	218	20
7	147	415	224	21
7	154	455	229	22
7	161	498	234	23
7	168	544	239	24
7	175	595	245	25
7	182	650	250	26
7	189	710	255	27
7	196	776	260	28
7	203	849	265	29
7	210	929	269	30
7	224	1117	278	32
7	238	1354	287	34
7	252	1661	296	36
7	266	2071	304	38
7	280	2649	312	40
7	294	3518	320	42
7	308	4971	328	44
7	322	7882	335	46
7	336	16626	343	48
8	8	8.7	27.4	1
8	16	19.3	63.2	2
8	24	31	85	3
8	32	43.6	102	4
8	40	57.4	117	5
8	48	72	130	6
8	56	88	142	7
8	64	105	153	8
8	72	124	164	9
8	80	143	173	10
8	88	164	182	11
8	96	187	191	12
8	104	211	199	13
8	112	236	207	14
8	120	264	215	15
8	128	293	222	16
8	136	324	230	17
8	144	358	237	18
8	152	394	243	19
8	160	433	250	20
8	168	475	256	21
8	176	521	263	22
8	184	570	269	23
8	192	623	275	24
8	200	681	281	25
8	208	744	286	26
8	216	812	292	27
8	224	888	297	28
8	232	971	303	29
8	240	1062	308	30
8	256	1277	318	32
8	272	1548	329	34
8	288	1899	338	36
8	304	2368	348	38
8	320	3028	357	40
8	336	4022	366	42
8	352	5682	375	44
8	368	9009	383	46
9	9	10	40	1
9	18	21.9	76	2
9	27	35	99	3
9	36	49.4	118	4
9	45	65	134	5
9	54	82	149	6
9	63	100	162	7
9	72	119	174	8
9	81	139	185	9
9	90	162	196	10
9	99	185	206	11
9	108	210	216	12
9	117	237	225	13
9	126	266	234	14
9	135	297	243	15
9	144	330	251	16
9	153	365	259	17
9	162	403	267	18
9	171	444	275	19
9	180	488	282	20
9	189	535	289	21
9	198	586	296	22
9	207	641	303	23
9	216	701	310	24
9	225	766	316	25
9	234	837	323	26
9	243	914	329	27
9	252	999	335	28
9	261	1092	341	29
9	270	1196	347	30
9	288	1438	359	32
9	306	1743	370	34
9	324	2137	381	36
9	342	2665	391	38
9	360	3407	402	40
9	378	4525	412	42
9	396	6393	422	44
10	10	11.2	50.5	1
10	20	24.5	87	2
10	30	39	112	3
10	40	55	133	4
10	50	72	151	5
10	60	91	167	6
10	70	111	181	7
10	80	132	195	8
10	90	155	207	9
10	100	180	219	10
10	110	206	230	11
10	120	234	241	12
10	130	264	251	13
10	140	296	261	14
10	150	330	271	15
10	160	367	280	16
10	170	406	289	17
10	180	448	297	18
10	190	494	306	19
10	200	543	314	20
10	210	595	322	21
10	220	652	330	22
10	230	713	337	23
10	240	780	345	24
10	250	852	352	25
10	260	930	359	26
10	270	1016	366	27
10	280	1111	373	28
10	290	1214	379	29
10	300	1329	383	30
10	320	1598	399	32
10	340	1937	411	34
10	360	2375	423	36
10	380	2961	435	38
10	400	3787	446	40
10	420	5029	458	42
10	440	7104	469	44

TABLE 4-2 (Cont'd.)

NETWORK B

The following is a computer solution for the Pi network when R_L equals 50 ohms.

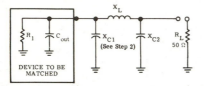

TO DESIGN A NETWORK USING THE TABLES

1. Define Q, in column one, as R_1/X_{C1}.

2. C_1 actual is equal to C_1 – parallel C_{out} of device to be matched.

3. This completes the network.

Q	X_{C1}	X_{C2}	X_L	R_1
1	1	5.03	5.47	1
1	2	7.14	8	2
1	3	8.79	10.03	3
1	4	10.21	11.8	4
1	5	11.47	13.4	5
1	10	16.67	20	10
1	15	21	25.35	15
1	20	25	30	20
1	25	28.87	34.15	25
1	30	32.73	37.91	30
1	35	36.69	41.35	35
1	40	40.82	44.49	40
1	45	45.23	47.37	45
1	50	50	50	50
1	55	55.28	52.37	55
1	60	61.24	54.49	60
1	65	68.14	56.35	65
1	70	76.38	57.91	70
1	75	86.6	59.15	75
1	80	100	60	80
1	85	119.02	60.35	85
1	90	150	60	90
2	0.5	3.17	3.56	1
2	1	4.49	5.25	2
2	1.5	5.51	6.64	3
2	2	6.38	7.87	4
2	2.5	7.14	9	5
2	5	10.21	13.8	10
2	7.5	12.63	17.87	15
2	10	14.74	21.56	20
2	12.5	16.67	25	25
2	15	18.46	28.25	30
2	17.5	20.17	31.35	35
2	20	21.82	34.33	40
2	22.5	23.43	37.21	45
2	25	25	40	50
2	27.5	26.55	42.71	55
2	30	28.1	45.35	60
2	32.5	29.64	47.93	65
2	35	31.18	50.45	70
2	37.5	32.73	52.91	75
2	40	34.3	55.32	80
2	42.5	35.89	57.69	85
2	45	37.5	60	90
2	47.5	39.14	62.27	95
2	50	40.82	64.49	100
2	62.5	50	75	125
2	75	61.24	84.49	150
2	87.5	76.38	92.91	175
2	100	100	100	200
2	112.5	150	105	225

Q	X_{C1}	X_{C2}	X_L	R_1
3	0.33	2.24	2.53	1
3	0.67	3.17	3.76	2
3	1	3.88	4.76	3
3	1.33	4.49	5.65	4
3	1.67	5.03	6.47	5
3	3.33	7.14	10	10
3	5	8.79	13.03	15
3	6.67	10.21	15.8	20
3	8.33	11.47	18.4	25
3	10	12.63	20.87	30
3	11.67	13.72	23.26	35
3	13.33	14.74	25.56	40
3	15	15.72	27.81	45
3	16.67	16.67	30	50
3	18.33	17.58	32.14	55
3	20	18.46	34.25	60
3	21.67	19.33	36.32	65
3	23.33	20.17	38.35	70
3	25	21	40.35	75
3	26.67	21.82	42.33	80
3	28.33	22.63	44.28	85
3	30	23.43	46.21	90
3	31.67	24.22	48.12	95
3	33.33	25	50	100
3	41.67	28.87	59.12	125
3	50	32.73	67.91	150
3	58.33	36.69	76.35	175
3	66.67	40.82	84.49	200
3	75	45.23	92.37	225
3	83.33	50	100	250
4	6.25	8.7	14.33	25
4	12.5	12.5	23.53	50
4	18.75	15.55	31.83	75
4	25	18.26	39.64	100
4	31.25	20.76	47.12	125
4	37.5	23.15	54.36	150
4	43.75	25.46	61.39	175
4	50	27.74	68.27	200
4	56.25	30	75	225
4	62.5	32.27	81.61	250
4	75	36.93	94.48	300
4	100	47.14	119.07	400
4	125	59.76	142.25	500
4	150	77.46	163.96	600
4	175	108.01	183.77	700
4	200	200	200	800
5	0.2	1.39	1.58	1
5	5	7	11.67	25
5	10	10	19.23	50
5	15	12.37	26.08	75

Q	X_{C1}	X_{C2}	X_L	R_1
5	20	14.43	32.55	100
5	25	16.31	38.78	125
5	30	18.06	44.82	150
5	35	19.72	50.72	175
5	40	21.32	56.5	200
5	45	22.87	62.18	225
5	50	24.4	67.78	250
5	60	27.39	78.76	300
5	80	33.33	100	400
5	100	39.53	120.48	500
5	120	46.29	140.31	600
5	140	54.01	159.54	700
5	160	63.25	178.17	800
5	180	75	196.15	900
5	200	91.29	213.37	1000
5	220	117.26	229.58	1100
5	240	173.21	244.09	1200
6	0.17	1.16	1.32	1
6	4.17	5.85	9.83	25
6	8.33	8.33	16.22	50
6	12.5	10.28	22.02	75
6	16.67	11.95	27.52	100
6	20.83	13.46	32.82	125
6	25	14.85	37.97	150
6	29.17	16.16	43.01	175
6	33.33	17.41	47.96	200
6	37.5	18.61	52.83	225
6	41.67	19.76	57.63	250
6	50	22	67.08	300
6	66.67	26.26	85.45	400
6	83.33	30.43	103.29	500
6	100	34.64	120.7	600
6	116.67	39.01	137.76	700
6	133.33	43.64	154.5	800
6	150	48.67	170.94	900
6	166.67	54.23	187.08	1000
6	183.33	60.55	202.93	1100
6	200	67.94	218.46	1200
6	216.67	76.87	233.66	1300
6	233.33	88.19	248.48	1400
6	250	103.51	262.83	1500
6	266.67	126.49	276.55	1600
6	283.33	168.33	289.32	1700
6	300	300	300	1800
7	0.14	1	1.14	1
7	3.57	5.03	8.47	25
7	7.14	7.14	14	50
7	10.71	8.79	19.03	75
7	14.29	10.21	23.8	100
7	17.86	11.47	28.4	125

TABLE 4-2 (Cont'd.)

Q	X_{C1}	X_{C2}	X_L	R_1
7	21.43	12.63	32.87	150
7	25	13.72	37.26	175
7	28.57	14.74	41.56	200
7	32.14	15.72	45.81	225
7	35.71	16.67	50	250
7	42.86	18.46	58.25	300
7	57.14	21.82	74.33	400
7	71.43	25	90	500
7	85.71	28.1	105.35	600
7	100	31.18	120.45	700
7	114.29	34.3	135.32	800
7	128.57	37.5	150	900
7	142.86	40.82	164.49	1000
7	171.43	48.04	192.98	1200
7	200	56.41	220.82	1400
7	228.57	66.67	248	1600
7	257.14	80.18	274.45	1800
7	285.71	100	300	2000
7	314.29	135.4	324.25	2200
7	342.86	244.95	345.8	2400
8	0.13	0.88	1	1
8	3.13	4.4	7.45	25
8	6.25	6.25	12.31	50
8	9.38	7.68	16.74	75
8	12.5	8.91	20.94	100
8	15.63	10	25	125
8	18.75	11	28.95	150
8	21.88	11.93	32.82	175
8	25	12.8	36.63	200
8	28.13	13.64	40.38	225
8	31.25	14.43	44.09	250
8	37.5	15.94	51.4	300
8	50	18.73	65.66	400
8	62.5	21.32	79.58	500
8	75	23.79	93.25	600
8	87.5	26.2	106.71	700
8	100	28.57	120	800
8	112.5	30.94	133.14	900
8	125	33.33	146.15	1000
8	150	38.25	171.82	1200
8	175	43.5	197.07	1400
8	200	49.24	221.92	1600
8	225	55.71	246.39	1800
8	250	63.25	270.48	2000
8	275	72.37	294.15	2200
8	300	84.02	317.36	2400
9	8.33	6.83	14.93	75
9	11.11	7.91	18.69	100
9	13.89	8.87	22.32	125
9	16.67	9.74	25.85	150
9	19.44	10.56	29.31	175
9	22.22	11.32	32.72	200
9	25	12.05	36.08	225
9	27.78	12.74	39.4	250
9	33.33	14.05	45.95	300
9	44.44	16.44	58.74	400
9	55.56	18.63	71.24	500
9	66.67	20.7	83.53	600
9	77.78	22.69	95.64	700
9	88.89	24.62	107.62	800
9	100	26.52	119.48	900
9	111.11	28.4	131.23	1000
9	133.33	32.16	154.46	1200
9	155.56	36	177.37	1400
9	177.78	40	200	1600
9	200	44.23	222.37	1800
9	222.22	48.8	244.5	2000
9	244.44	53.8	266.4	2200
9	266.67	59.41	288.05	2400

Q	X_{C1}	X_{C2}	X_L	R_1
10	0.1	0.7	0.8	1
10	5	5	9.9	50
10	10	7.11	16.87	100
10	15	8.75	23.34	150
10	20	10.15	29.55	200
10	25	11.41	35.6	250
10	30	12.57	41.52	300
10	40	14.66	53.11	400
10	50	16.57	64.44	500
10	60	18.36	75.58	600
10	70	20.06	86.58	700
10	80	21.69	97.46	800
10	90	23.28	108.24	900
10	100	24.85	118.94	1000
10	120	27.91	140.09	1200
10	140	30.97	161	1400
10	160	34.05	181.68	1600
10	180	37.21	202.17	1800
10	200	40.49	222.47	2000
10	220	43.93	242.61	2200
10	240	47.58	262.59	2400
12	25	10.39	34.79	300
12	33.33	12.08	44.52	400
12	41.67	13.61	54.05	500
12	50	15.02	63.43	600
12	58.33	16.35	72.7	700
12	66.67	17.61	81.87	800
12	75	18.82	90.97	900
12	83.33	20	100	1000
12	100	22.27	117.89	1200
12	116.67	24.46	135.6	1400
12	133.33	26.61	153.15	1600
12	150	28.73	170.57	1800
12	166.67	30.86	187.86	2000
12	183.33	33	205.06	2200
12	200	35.17	222.15	2400
12	216.67	37.39	239.16	2600
12	233.33	39.66	256.07	2800
12	250	42.01	272.9	3000
12	291.67	48.3	314.64	3500
12	333.33	55.47	355.9	4000
12	375	63.96	396.67	4500
12	416.67	74.54	436.92	5000
12	458.33	88.64	476.57	5500
12	500	109.54	515.44	6000
14	21.43	8.86	29.91	300
14	28.57	10.29	38.3	400
14	35.71	11.56	46.51	500
14	42.86	12.73	54.6	600
14	50	13.83	62.59	700
14	57.14	14.87	70.51	800
14	64.29	15.86	78.37	900
14	71.43	16.81	86.17	1000
14	85.71	18.62	101.63	1200
14	100	20.35	116.95	1400
14	114.29	22.02	132.15	1600
14	128.57	23.64	147.24	1800
14	142.86	25.24	162.25	2000
14	157.14	26.81	177.17	2200
14	171.43	28.38	192.02	2400
14	185.71	29.94	206.81	2600
14	200	31.51	221.54	2800
14	214.29	33.09	236.21	3000
14	250	37.12	272.66	3500
14	285.71	41.34	308.82	4000
14	321.43	45.86	344.7	4500
14	357.14	50.77	380.33	5000
14	392.86	56.22	415.69	5500
14	428.57	62.42	450.79	6000

Q	X_{C1}	X_{C2}	X_L	R_1
16	18.75	7.73	26.23	300
16	25	8.96	33.59	400
16	31.25	10.06	40.8	500
16	37.5	11.07	47.9	600
16	43.75	12	54.93	700
16	50	12.88	61.89	800
16	56.25	13.72	68.79	900
16	62.5	14.52	75.65	1000
16	75	16.05	89.26	1200
16	87.5	17.48	102.74	1400
16	100	18.86	116.12	1600
16	112.5	20.18	129.42	1800
16	125	21.47	142.64	2000
16	137.5	22.73	155.8	2200
16	150	23.96	168.9	2400
16	162.5	25.18	181.95	2600
16	175	26.39	194.96	2800
16	187.5	27.59	207.92	3000
16	218.75	30.59	240.16	3500
16	250	33.61	272.18	4000
16	281.25	36.71	304.01	4500
16	312.5	39.9	335.66	5000
16	343.75	43.25	367.15	5500
16	375	46.8	398.49	6000
18	16.67	6.86	23.35	300
18	22.22	7.94	29.9	400
18	27.78	8.91	36.33	500
18	33.33	9.79	42.66	600
18	38.89	10.61	48.92	700
18	44.44	11.38	55.13	800
18	50	12.11	61.28	900
18	55.56	12.8	67.4	1000
18	66.67	14.12	79.54	1200
18	77.78	15.35	91.57	1400
18	88.89	16.52	103.51	1600
18	100	17.65	115.38	1800
18	111.11	18.73	127.2	2000
18	122.22	19.79	138.95	2200
18	133.33	20.81	150.66	2400
18	144.44	21.82	162.33	2600
18	155.56	22.81	173.96	2800
18	166.67	23.79	185.55	3000
18	194.44	26.2	214.4	3500
18	222.22	28.57	243.08	4000
18	250	30.94	271.6	4500
18	277.78	33.33	300	5000
18	305.56	35.76	328.27	5500
18	333.33	38.25	356.44	6000
20	15	6.16	21.03	300
20	20	7.13	26.94	400
20	25	8	32.73	500
20	30	8.78	38.44	600
20	35	9.51	44.09	700
20	40	10.19	49.69	800
20	45	10.84	55.24	900
20	50	11.46	60.76	1000
20	60	12.62	71.71	1200
20	70	13.7	82.57	1400
20	80	14.72	93.35	1600
20	90	15.7	104.07	1800
20	100	16.64	114.73	2000
20	110	17.55	125.35	2200
20	120	18.44	135.93	2400
20	130	19.3	146.47	2600
20	140	20.14	156.98	2800
20	150	20.97	167.46	3000
20	175	22.99	193.54	3500
20	200	24.96	219.48	4000
20	225	26.9	245.3	4500
20	250	28.82	271.01	5000
20	275	30.74	296.62	5500
20	300	32.67	322.15	6000

TABLE 4-2 (Cont'd.)

NETWORK C_1

The following is a computer solution for an RF matching network. This computer solution is applicable for two forms of matching networks.

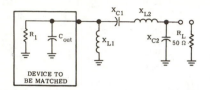

DEVICE TO
BE MATCHED

TO DESIGN A NETWORK USING THE TABLES

1. $X_{L1} = X_{C\,out}$.
2. Define Q, in column one, as X_{C1}/R_1.
3. All network values can now be read from the charts in terms of reactance.
4. This completes network C_1.

NETWORK C_2

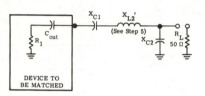

DEVICE TO
BE MATCHED

TO DESIGN A NETWORK USING THE TABLES

1. L_1 is not used in this network.
2. Transform the impedance of the device to be matched to series form $(R_1 + jX_{C\,out})$.
3. Define Q, in column one, as X_{C1}/R_1.
4. For a desired Q, find the R_s to be matched in the R_1 column and read the reactive value of the components.
5. X_{L2}' is equal to the quantity X_{L2} obtained from the tables plus $|X_{C\,out}|$.
6. This completes network C_2.

Q	X_{C1}	X_{C2}	X_{L2}	R_1	Q	X_{C1}	X_{C2}	X_{L2}	R_1	Q	X_{C1}	X_{C2}	X_{L2}	R_1
1	1	7.14	8	1	1	38	88.98	59.35	38	2	54	54.17	78.92	27
1	2	10.21	11.8	2	1	40	100	60	40	2	56	56.41	80.82	28
1	3	12.63	14.87	3	1	42	114.56	60.33	42	2	58	58.76	82.68	29
1	4	14.74	17.56	4	1	44	135.4	60.25	44	2	60	61.24	84.49	30
1	5	16.67	20	5	1	46	169.56	59.56	46	2	64	66.67	88	32
1	6	18.46	22.25	6	1	48	244.95	57.8	48	2	68	72.89	91.32	34
1	7	20.17	24.35	7						2	72	80.18	94.45	36
1	8	21.82	26.33	8	2	2	7.14	9	1	2	76	88.98	97.35	38
1	9	23.43	28.21	9	2	4	10.21	13.8	2	2	80	100	100	40
1	10	25	30	10	2	6	12.63	17.87	3	2	84	114.56	102.33	42
1	11	26.55	31.81	11	2	8	14.74	21.56	4	2	88	135.4	104.25	44
1	12	28.1	33.35	12	2	10	16.67	25	5	2	92	169.56	105.56	46
1	13	29.64	34.93	13	2	12	-18.46	28.25	6	2	96	244.95	105.8	48
1	14	31.13	36.45	14	2	14	20.17	31.35	7					
1	15	32.73	37.91	15	2	16	21.82	34.33	8	3	3	7.14	10	1
1	16	34.3	39.32	16	2	18	23.43	37.21	9	3	6	10.21	15.8	2
1	17	35.89	40.69	17	2	20	25	40	10	3	9	12.63	20.87	3
1	18	37.5	42	18	2	22	26.55	42.71	11	3	12	14.74	25.56	4
1	19	39.14	43.27	19	2	24	28.1	45.35	12	3	15	16.67	30	5
1	20	40.82	44.49	20	2	26	29.64	47.93	13	3	18	18.46	34.25	6
1	21	42.55	45.68	21	2	28	31.18	50.45	14	3	21	20.17	38.35	7
1	22	44.32	46.82	22	2	30	32.73	52.91	15	3	24	21.82	42.33	8
1	23	46.15	47.92	23	2	32	34.3	55.32	16	3	27	23.43	46.21	9
1	24	48.04	48.98	24	2	34	35.89	57.69	17	3	30	25	50	10
1	25	50	50	25	2	36	37.5	60	18	3	33	26.55	53.71	11
1	26	52.04	50.98	26	2	38	39.14	62.27	19	3	36	28.1	57.35	12
1	27	54.17	51.92	27	2	40	40.82	64.49	20	3	39	29.64	60.98	13
1	28	56.41	52.82	28	2	42	42.55	66.68	21	3	42	31.18	64.45	14
1	29	58.76	53.68	29	2	44	44.32	68.82	22	3	45	32.73	67.91	15
1	30	61.24	54.49	30	2	46	46.15	70.92	23	3	48	34.3	71.32	16
1	32	66.67	56	32	2	48	48.04	72.98	24	3	51	35.89	74.69	17
1	34	72.89	57.32	34	2	50	50	75	25	3	54	37.5	78	18
1	36	80.18	58.45	36	2	52	52.04	76.98	26	3	57	39.14	81.27	19

TABLE 4-2 (Cont'd.)

Q	X_{C1}	X_{C2}	X_{L2}	R_1
3	60	40.82	84.49	20
3	63	42.55	87.68	21
3	66	44.32	90.82	22
3	69	46.15	93.93	23
3	72	48.04	96.98	24
3	75	50	100	25
3	78	52.04	102.98	26
3	81	54.17	105.92	27
3	84	56.41	108.82	28
3	87	58.76	111.68	29
3	90	61.24	114.49	30
3	96	66.67	120	32
3	102	72.89	125.32	34
3	108	80.18	130.45	36
3	114	88.98	135.35	38
3	120	100	140	40
3	126	114.56	144.33	42
3	132	135.4	148.25	44
3	138	169.56	151.56	46
3	144	244.95	153.8	48
4	4	7.14	11	1
4	8	10.21	17.8	2
4	12	12.63	23.87	3
4	16	14.74	29.56	4
4	20	16.67	35	5
4	24	18.46	40.25	6
4	28	20.17	45.35	7
4	32	21.82	50.33	8
4	36	23.43	55.21	9
4	40	25	60	10
4	44	26.55	64.71	11
4	48	28.1	69.35	12
4	52	29.64	73.93	13
4	56	31.18	78.45	14
4	60	32.73	82.91	15
4	64	34.3	87.32	16
4	68	35.89	91.69	17
4	72	37.5	96	18
4	76	39.14	100.27	19
4	80	40.82	104.49	20
4	84	42.55	108.68	21
4	88	44.32	112.82	22
4	92	46.15	116.92	23
4	96	48.04	120.98	24
4	100	50	125	25
4	104	52.04	128.98	26
4	108	54.17	132.92	27
4	112	56.41	136.82	28
4	116	58.76	140.68	29
4	120	61.24	144.49	30
4	128	66.67	152	32
4	136	72.89	159.32	34
4	144	80.18	166.45	36
4	152	88.98	173.35	38
4	160	100	180	40
4	168	114.56	186.33	42
4	176	135.4	192.25	44
4	184	169.56	197.56	46
4	192	244.95	201.8	48
5	5	7.14	12	1
5	10	10.21	19.8	2
5	15	12.63	26.87	3
5	20	14.74	33.56	4
5	25	16.67	40	5
5	30	18.46	46.25	6
5	35	20.17	52.35	7
5	40	21.82	58.33	8
5	45	23.43	64.21	9
5	50	25	70	10
5	55	26.55	75.71	11

Q	X_{C1}	X_{C2}	X_{L2}	R_1
5	60	28.1	81.35	12
5	65	29.64	86.93	13
5	70	31.18	92.45	14
5	75	32.73	97.91	15
5	80	34.3	103.32	16
5	85	35.89	108.69	17
5	90	37.5	114	18
5	95	39.14	119.27	19
5	100	40.82	124.49	20
5	105	42.55	129.68	21
5	110	44.32	134.82	22
5	115	46.15	139.92	23
5	120	48.04	144.98	24
5	125	50	150	25
5	130	52.04	154.98	26
5	135	54.17	159.92	27
5	140	56.41	164.82	28
5	145	58.76	169.68	29
5	150	61.24	174.49	30
5	160	66.67	184	32
5	170	72.89	193.32	34
5	180	80.18	202.45	36
5	190	88.98	211.35	38
5	200	100	220	40
5	210	114.56	228.33	42
5	220	135.4	236.25	44
5	230	169.56	243.56	46
5	240	244.95	249.8	48
6	6	7.14	13	1
6	12	10.21	21.8	2
6	18	12.63	29.87	3
6	24	14.74	37.56	4
6	30	16.67	45	5
6	36	18.46	52.25	6
6	42	20.17	59.35	7
6	48	21.82	66.33	8
6	54	23.43	73.21	9
6	60	25	80	10
6	66	26.55	86.71	11
6	72	28.1	93.35	12
6	78	29.64	99.93	13
6	84	31.18	106.45	14
6	90	32.73	112.91	15
6	96	34.3	119.32	16
6	102	35.89	125.69	17
6	108	37.5	132	18
6	114	39.14	138.27	19
6	120	40.82	144.49	20
6	126	42.55	150.68	21
6	132	44.32	156.82	22
6	138	46.15	162.92	23
6	144	48.04	168.98	24
6	150	50	175	25
6	156	52.04	180.98	26
6	162	54.17	186.92	27
6	168	56.41	192.82	28
6	174	58.76	198.68	29
6	180	61.24	204.49	30
6	192	66.67	216	32
6	204	72.89	227.32	34
6	216	80.18	238.45	36
6	228	88.98	249.35	38
6	240	100	260	40
6	252	114.56	270.33	42
6	264	135.4	280.25	44
6	276	169.56	289.56	46
6	288	244.95	297.8	48
7	7	7.14	14	1
7	14	10.21	23.8	2
7	21	12.63	32.87	3

Q	X_{C1}	X_{C2}	X_{L2}	R_1
7	28	14.74	41.56	4
7	35	16.67	50	5
7	42	18.46	58.25	6
7	49	20.17	66.35	7
7	56	21.82	74.33	8
7	63	23.43	82.21	9
7	70	25	90	10
7	77	26.55	97.71	11
7	84	28.1	105.35	12
7	91	29.64	112.93	13
7	98	31.18	120.45	14
7	105	32.73	127.91	15
7	112	34.3	135.32	16
7	119	35.89	142.69	17
7	126	37.5	150	18
7	133	39.14	157.27	19
7	140	40.82	164.49	20
7	147	42.55	171.68	21
7	154	44.32	178.82	22
7	161	46.15	185.92	23
7	168	48.04	192.98	24
7	175	50	200	25
7	182	52.04	206.98	26
7	189	54.17	213.92	27
7	196	56.41	220.82	28
7	203	58.76	227.68	29
7	210	61.24	234.49	30
7	224	66.67	248	32
7	238	72.89	261.32	34
7	252	80.18	274.45	36
7	266	88.98	287.35	38
7	280	100	300	40
7	294	114.56	312.33	42
7	308	135.4	324.25	44
7	322	169.56	335.56	46
7	336	244.95	345.8	48
8	8	7.14	15	1
8	16	10.21	25.8	2
8	24	12.63	35.87	3
8	32	14.74	45.56	4
8	40	16.67	55	5
8	48	18.46	64.25	6
8	56	20.17	73.35	7
8	64	21.82	82.33	8
8	72	23.43	91.21	9
8	80	25	100	10
8	88	26.55	108.71	11
8	96	28.1	117.35	12
8	104	29.64	125.93	13
8	112	31.18	134.45	14
8	120	32.73	142.91	15
8	128	34.3	151.32	16
8	136	35.89	159.69	17
8	144	37.5	168	18
8	152	39.14	176.27	19
8	160	40.82	184.49	20
8	168	42.55	192.68	21
8	176	44.32	200.82	22
8	184	46.15	208.92	23
8	192	48.04	216.98	24
8	200	50	225	25
8	208	52.04	232.98	26
8	216	54.17	240.92	27
8	224	56.41	248.82	28
8	232	58.76	256.68	29
8	240	61.24	264.49	30
8	256	66.67	280	32
8	272	72.89	295.32	34
8	288	80.18	310.45	36
8	304	88.98	325.35	38

TABLE 4-2 (Cont'd.)

Q	X_{C1}	X_{C2}	X_{L2}	R_1
8	320	100	340	40
8	336	114.56	354.33	42
8	352	135.4	368.25	44
8	368	.169.56	381.56	46
8	384	244.95	393.8	48
9	9	7.14	16	1
9	18	10.21	27.8	2
9	27	12.63	38.87	3
9	36	14.74	49.56	4
9	45	16.67	60	5
9	54	18.46	70.25	6
9	63	20.17	80.35	7
9	72	21.82	90.33	8
9	81	23.43	100.21	9
9	90	25	110	10
9	99	26.55	119.71	11
9	108	28.1	129.35	12
9	117	29.64	138.93	13
9	126	31.18	148.45	14
9	135	32.73	157.91	15
9	144	34.3	167.32	16
9	153	35.89	176.69	17
9	162	37.5	186	18
9	171	39.17	195.27	19
9	180	40.82	204.49	20
9	189	42.55	213.68	21
9	198	44.32	222.82	22
9	207	46.15	231.92	23

Q	X_{C1}	X_{C2}	X_{L2}	R_1
9	414	169.56	427.56	46
9	432	244.95	441.8	48
9	216	48.04	240.98	24
9	225	50	250	25
9	234	52.04	258.98	26
9	243	54.17	267.92	27
9	252	56.41	276.82	28
9	261	58.76	285.88	29
9	270	61.24	294.49	30
9	288	66.67	312	32
9	306	72.89	329.32	34
9	324	80.18	346.45	36
9	342	88.98	363.35	38
9	360	100	380	40
9	378	114.56	396.33	42
9	396	135.4	412.25	44
10	10	7.14	17	1
10	20	10.21	29.8	2
10	30	12.63	41.87	3
10	40	14.74	53.56	4
10	50	16.67	65	5
10	60	18.46	76.25	6
10	70	20.17	87.35	7
10	80	21.82	98.33	8
10	90	23.43	109.21	9
10	100	25	120	10
10	110	26.55	130.71	11

Q	X_{C1}	X_{C2}	X_{L2}	R_1
10	120	28.1	141.35	12
10	130	29.64	151.93	13
10	140	31.18	162.45	14
10	150	32.73	172.91	15
10	160	34.3	183.32	16
10	170	35.89	193.69	17
10	180	37.5	204	18
10	190	39.14	214.27	19
10	200	40.82	224.49	20
10	210	42.55	234.68	21
10	220	44.32	244.82	22
10	230	46.15	254.92	23
10	240	48.04	264.98	24
10	250	50	275	25
10	260	52.04	284.98	26
10	270	54.17	294.92	27
10	280	56.41	304.82	28
10	290	58.76	314.68	29
10	300	61.24	324.49	30
10	320	66.67	344	32
10	340	72.89	363.32	34
10	360	80.18	382.45	36
10	380	88.98	401.35	38
10	400	100	420	40
10	420	114.56	438.33	42
10	440	135.4	456.25	44
10	460	169.56	473.56	46
10	480	244.95	489.8	48

NETWORK D

The following is a computer solution for an RF "Tee" matching network.

Tuning is accomplished by using a variable capacitor for C_1. Variable matching may also be accomplished by increasing X_{L2} and adding an equal amount of X_C in series in the form of a variable capacitor.

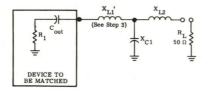

TO DESIGN A NETWORK USING THE TABLES

1. Define Q, in column one, as X_{L1}/R_1.

2. For an R_1 to be matched and a desired Q, read the reactances of the network components from the charts.

3. X_{L1}' is equal to the quantity X_{L1} obtained from the tables plus $|X_{C_{out}}|$.

4. This completes the network.

Q	X_{L1}	X_{L2}	X_{C1}	R_1
1	26	10	43.33	26
1	27	14.14	42.09	27
1	28	17.32	41.59	28
1	29	20	41.43	29
1	30	22.36	41.46	30
1	32	26.46	41.85	32
1	34	30	42.5	34
1	36	33.17	43.29	36
1	38	36.06	44.16	38
1	40	38.72	45.08	40
1	42	41.23	46.04	42
1	44	43.59	47.01	44
1	46	45.83	48	46
1	48	47.96	49	48
1	50	50	50	50
1	55	54.77	52.49	55
1	60	59.16	54.96	60
1	65	63.25	57.4	65
1	70	67.08	69.79	70
1	75	70.71	62.13	75
1	80	74.16	64.43	80
1	85	77.46	66.69	85
1	90	80.62	68.9	90
1	95	83.67	71.07	95
1	100	86.6	73.21	100
1	125	100	83.33	125
1	150	111.8	92.71	150

Q	X_{L1}	X_{L2}	X_{C1}	R_1
1	175	122.47	101.46	175
1	200	132.29	109.72	200
1	225	141.42	117.54	225
1	250	150	125	250
1	275	158.11	132.14	275
1	300	165.83	139	300
2	22	15.81	23.75	11
2	24	22.36	24.52	12
2	26	27.39	25.51	13
2	28	31.62	26.59	14
2	30	35.36	27.7	15
2	32	38.73	28.83	16
2	34	41.83	29.96	17
2	36	44.72	31.09	18
2	38	47.43	32.22	19
2	40	50	33.33	20
2	42	52.44	34.44	21
2	44	54.77	35.54	22
2	46	57.01	36.62	23
2	48	59.16	37.7	24
2	50	61.24	38.76	25
2	52	63.25	39.82	26
2	54	65.19	40.86	27
2	56	67.08	41.9	28
2	58	68.92	42.92	29
2	60	70.71	43.93	30
2	64	74.16	45.93	32

Q	X_{L1}	X_{L2}	X_{C1}	R_1
2	68	77.46	47.9	34
2	72	80.62	49.83	36
2	76	83.67	51.72	38
2	80	86.6	53.59	40
2	84	89.44	55.43	42
2	88	92.2	57.23	44
2	92	94.87	59.01	46
2	96	97.47	60.77	48
2	100	100	62.5	50
2	110	106.07	66.73	55
2	120	111.8	70.82	60
2	130	117.26	74.8	65
2	140	122.47	78.66	70
2	150	127.48	82.43	75
2	160	132.29	86.1	80
2	170	136.93	89.69	85
2	180	141.42	93.2	90
2	190	145.77	96.63	95
2	200	150	100	100
2	250	169.56	115.93	125
2	300	187.08	130.62	150
2	350	203.1	144.34	175
2	400	217.94	157.26	200
2	450	231.84	169.51	225
2	500	244.95	181.19	250
2	550	257.39	192.37	275
2	600	269.26	203.11	300

TABLE 4-2 (Cont'd.)

Q	X_{L1}	X_{L2}	X_{C1}	R_1	Q	X_{L1}	X_{L2}	X_{C1}	R_1	Q	X_{L1}	X_{L2}	X_{C1}	R_1
3	18	22.36	17.41	6	4	112	145.95	68.8	28	5	625	400	250	125
3	21	31.62	19.27	7	4	116	148.83	70.67	29	5	750	438.75	283.12	150
3	24	38.73	21.19	8	4	120	151.66	72.51	30	5	875	474.34	314.08	175
3	27	44.72	23.11	9	4	128	157.16	76.16	32	5	1000	507.44	343.26	200
3	30	50	25	10	4	136	162.48	79.73	34	5	1125	538.52	670.95	225
3	33	54.77	26.86	11	4	144	167.63	83.24	36	5	1250	567.89	397.36	250
3	36	59.16	28.69	12	4	152	172.63	86.68	38	5	1375	595.82	422.67	275
3	39	63.25	30.48	13	4	160	177.48	90.07	40	5	1500	622.49	446.99	300
3	42	67.08	32.25	14	4	168	182.21	93.4	42	6	12	34.64	11.06	2
3	45	70.71	33.98	15	4	176	186.82	96.69	44	6	18	55.23	15.62	3
3	48	74.16	35.69	16	4	184	191.31	99.92	46	6	24	70	20	4
3	51	77.46	37.37	17	4	192	195.7	103.11	48	6	30	82.16	24.2	5
3	54	80.62	39.02	18	4	200	200	106.25	50	6	36	92.74	28.26	6
3	57	83.67	40.66	19	4	220	210.36	113.93	55	6	42	102.23	32.2	7
3	60	86.6	42.26	20	4	240	220.23	121.36	60	6	48	110.91	36.02	8
3	63	89.44	43.85	21	4	260	229.67	128.59	65	6	54	118.95	39.74	9
3	66	92.2	45.42	22	4	280	238.75	135.61	70	6	60	126.49	43.38	10
3	69	94.87	46.96	23	4	300	247.49	142.46	75	6	66	133.6	46.93	11
3	72	97.47	48.49	24	4	320	255.93	148.15	80	6	72	140.36	50.41	12
3	75	100	50	25	4	340	264.1	155.68	85	6	78	146.8	53.83	13
3	78	102.47	51.49	26	4	360	272.03	162.07	90	6	84	152.97	57.18	14
3	81	104.88	52.97	27	4	380	279.73	168.32	95	6	90	158.74	60.47	15
3	84	107.24	54.42	28	4	400	287.23	174.46	100	6	96	164.62	63.71	16
3	87	109.54	55.87	29	4	500	322.1	203.5	125	6	102	170.15	66.89	17
3	90	111.8	57.29	30	4	600	353.55	230.33	150	6	108	175.5	70.03	18
3	96	116.19	60.11	32	4	700	382.43	255.4	175	6	114	180.69	73.12	19
3	102	120.42	62.87	34	4	800	409.27	279.02	200	6	120	185.74	76.17	20
3	108	124.5	65.57	36	4	900	434.45	301.44	225	6	126	190.66	79.18	21
3	114	128.45	68.23	38	4	1000	458.26	322.82	250	6	132	195.45	82.15	22
3	120	132.29	70.85	40	4	1100	480.88	343.3	275	6	138	200.12	85.08	23
3	126	136.01	73.42	42	4	1200	502.49	362.99	300	6	144	204.69	87.97	24
3	132	139.64	75.96	44						6	150	209.17	90.83	25
3	138	143.18	78.45	46	5	10	10	10	2	6	156	213.54	93.66	26
3	144	146.63	80.91	48	5	15	37.42	13.57	3	6	162	217.83	96.46	27
3	150	150	83.33	50	5	20	51.96	17.22	4	6	168	222.04	99.23	28
3	165	158.11	89.25	55	5	25	63.25	20.75	5	6	174	226.16	101.96	29
3	180	165.83	94.99	60	5	30	72.8	24.16	6	6	180	230.22	104.67	30
3	195	173.21	100.56	65	5	35	81.24	27.47	7	6	192	238.12	110.01	32
3	210	180.28	105.97	70	5	40	88.88	30.69	8	6	204	245.76	115.25	34
3	225	187.08	111.25	75	5	45	95.92	33.82	9	6	216	253.18	120.39	36
3	240	193.65	116.4	80	5	50	102.47	36.88	10	6	228	260.38	125.45	38
3	255	200	121.43	85	5	55	108.63	39.87	11	6	240	267.39	130.42	40
3	270	206.16	126.35	90	5	60	114.46	42.8	12	6	252	274.23	135.31	42
3	285	212.13	131.17	95	5	65	120	45.68	13	6	264	280.89	140.13	44
3	300	217.94	135.89	100	5	70	125.3	48.49	14	6	276	287.4	144.88	46
3	375	244.95	158.25	125	5	75	130.38	51.26	15	6	288	293.77	149.55	48
3	450	269.26	178.89	150	5	80	135.28	53.99	16	6	300	300	154.17	50
3	525	291.55	198.17	175	5	85	140	56.67	17	6	330	315.04	165.44	55
3	600	312.25	216.33	200	5	90	144.57	59.31	18	6	360	329.39	176.36	60
3	675	331.66	233.57	225	5	95	149	61.91	19	6	390	343.15	186.97	65
3	750	350	250	250	5	100	153.3	64.47	20	6	420	356.37	197.3	70
3	825	367.42	265.74	275	5	105	157.48	67	21	6	450	369.12	207.36	75
3	900	384.06	280.87	300	5	110	161.55	69.49	22	6	480	381.44	217.19	80
					5	115	165.53	71.96	23	6	510	393.38	226.79	85
					5	120	169.41	74.39	24	6	540	404.97	236.18	90
					5	125	173.21	76.79	25	6	570	416.23	245.38	95
4	12	7.07	12.31	3	5	130	176.92	79.17	26	6	600	427.2	254.4	100
4	16	30	14.78	4	5	135	180.55	81.52	27	6	750	478.28	297.13	125
4	20	41.83	17.57	5	5	140	184.12	83.85	28	6	900	524.4	336.61	150
4	24	50.99	20.32	6	5	145	187.62	86.15	29	6	1050	566.79	373.5	175
4	28	58.74	23	7	5	150	191.05	88.43	30	6	1200	606.22	408.29	200
4	32	65.57	25.6	8	5	160	197.74	92.91	32	6	1350	643.23	441.3	225
4	36	71.76	28.15	9	5	170	204.21	97.31	34	6	1500	678.23	472.79	250
4	40	77.46	30.64	10	5	180	210.48	101.63	36	6	1650	711.51	502.96	275
4	44	82.76	33.07	11	5	190	216.56	105.88	38	6	1800	743.3	531.96	300
4	48	87.75	35.45	12	5	200	222.49	110.06	40					
4	52	92.47	37.78	13	5	210	228.25	114.17	42	7	14	50	12.5	2
4	56	96.95	40.07	14	5	220	233.88	118.21	44	7	21	70.71	17.83	3
4	60	101.24	42.32	15	5	230	239.37	122.2	46	7	28	86.6	22.9	4
4	64	105.36	44.54	16	5	240	244.74	126.13	48	7	35	100	27.78	5
4	68	109.32	46.72	17	5	250	260	130	50	7	42	111.8	32.48	6
4	72	113.14	48.86	18	5	275	262.68	139.46	55	7	49	122.47	37.04	7
4	76	116.83	50.97	19	5	300	274.77	148.64	60	7	56	132.29	41.47	8
4	80	120.42	53.06	20	5	325	286.36	157.54	65	7	63	141.42	45.79	9
4	84	123.9	55.11	21	5	350	297.49	166.21	70	7	70	150	50	10
4	88	127.28	57.14	22	5	375	308.22	174.66	75	7	77	158.11	54.12	11
4	92	130.58	59.14	23	5	400	318.59	182.91	80	7	84	165.83	58.16	12
4	96	133.79	61.12	24	5	425	328.63	190.97	85	7	91	173.21	62.12	13
4	100	136.93	63.07	25	5	450	338.38	198.85	90	7	98	180.28	66	14
4	104	140	65	26	5	475	347.85	206.57	95	7	105	187.08	69.82	15
4	108	143	66.91	27	5	500	357.07	214.14	100	7	112	193.65	73.58	16

TABLE 4-2 (Cont'd.)

Q	X_{L1}	X_{L2}	X_{C1}	R_1
7	119	200	77.27	17
7	126	206.16	80.91	18
7	133	212.13	84.5	19
7	140	217.94	88.04	20
7	147	223.61	91.53	21
7	154	229.13	94.97	22
7	161	234.52	98.37	23
7	168	239.79	101.73	24
7	175	244.95	105.05	25
7	182	250	108.33	26
7	189	254.95	111.58	27
7	196	259.81	114.79	28
7	203	264.58	117.97	29
7	210	269.26	121.11	30
7	224	278.39	127.31	32
7	238	287.23	133.39	34
7	252	295.8	139.36	36
7	266	304.14	145.23	38
7	280	312.25	151	40
7	294	320.16	156.68	42
7	308	327.87	162.27	44
7	322	335.41	167.78	46
7	336	342.78	173.21	48
7	350	350	178.57	50
7	385	367.42	191.66	55
7	420	384.06	204.34	60
7	455	400	216.67	65
7	490	415.33	228.66	70
7	525	430.12	240.35	75
7	560	444.41	251.76	80
7	595	458.86	262.91	85
7	630	471.7	273.82	90
7	665	484.77	284.51	95
7	700	497.49	294.99	100
7	875	556.78	344.63	125
7	1050	610.33	390.49	150
7	1225	659.55	433.36	175
7	1400	705.34	473.78	200
7	1575	748.33	512.14	225
7	1750	788.99	548.73	250
7	1925	827.65	583.79	275
7	2100	864.58	617.5	300
8	8	27.39	7.6	1
8	16	63.25	14.03	2
8	24	85.15	20.1	3
8	32	102.47	25.87	4
8	40	117.26	31.42	5
8	48	130.38	36.77	6
8	56	142.3	41.95	7
8	64	153.3	46.99	8
8	72	163.55	51.9	9
8	80	173.21	56.7	10
8	88	182.35	61.39	11
8	96	191.05	65.98	12
8	104	199.37	70.49	13
8	112	207.36	74.91	14
8	120	215.06	79.26	15
8	128	222.49	83.54	16
8	136	229.67	87.74	17
8	144	236.64	91.89	18
8	152	243.41	95.97	19
8	160	250	100	20
8	168	256.42	103.97	21
8	176	262.68	107.9	22
8	184	268.79	111.77	23
8	192	274.77	115.59	24
8	200	280.62	119.38	25
8	208	286.36	123.11	26
8	216	291.98	126.81	27
8	224	297.49	130.47	28
8	232	302.9	134.09	29
8	240	308.22	137.67	30

Q	X_{L1}	X_{L2}	X_{C1}	R_1
8	256	318.59	144.73	32
8	272	328.63	151.65	34
8	288	338.38	158.46	36
8	304	347.85	165.14	38
8	320	357.07	171.71	40
8	336	366.06	178.18	42
8	352	374.83	184.56	44
8	368	383.41	190.83	46
8	384	391.79	197.02	48
8	400	400	203.13	50
8	440	419.82	218.04	55
8	480	438.75	232.49	60
8	520	456.89	246.53	65
8	560	474.34	260.2	70
8	600	491.17	273.52	75
8	640	507.44	286.52	80
8	680	523.21	299.23	85
8	720	538.52	311.66	90
8	760	553.4	323.84	95
8	800	567.89	335.78	100
8	1000	635.41	392.36	125
8	1200	696.42	444.63	150
8	1400	752.5	493.49	175
8	1600	804.67	539.57	200
8	1800	853.67	583.29	225
8	2000	900	625	250
8	2200	944.06	664.96	275
8	2400	986.15	703.38	300
9	9	40	8.37	1
9	18	75.5	15.6	2
9	27	98.99	22.4	3
9	36	117.9	28.88	4
9	45	134.16	35.09	5
9	54	148.66	41.09	6
9	63	161.86	46.91	7
9	72	174.07	52.56	8
9	81	185.47	58.07	9
9	90	196.21	63.45	10
9	99	206.4	68.71	11
9	108	216.1	73.86	12
9	117	225.39	78.92	13
9	126	234.31	83.88	14
9	135	242.9	88.76	15
9	144	251.2	93.55	16
9	153	259.23	98.28	17
9	162	267.02	102.93	18
9	171	274.59	107.51	19
9	180	281.96	112.03	20
9	189	289.14	116.49	21
9	198	296.14	120.89	22
9	207	302.99	125.23	23
9	216	309.68	129.53	24
9	225	316.23	133.77	25
9	234	322.65	137.97	26
9	243	328.94	142.12	27
9	252	335.11	146.22	28
9	261	341.17	150.28	29
9	270	347.13	154.3	30
9	288	358.75	162.23	32
9	306	370	170	34
9	324	380.92	177.63	36
9	342	391.54	185.14	38
9	360	401.87	192.52	40
9	378	411.95	199.78	42
9	396	421.78	206.93	44
9	414	431.39	213.98	46
9	432	440.79	220.93	48
9	450	450	227.78	50
9	495	472.23	244.52	55
9	540	493.46	260.74	60
9	585	513.81	276.51	65
9	630	533.39	291.85	70

Q	X_{L1}	X_{L2}	X_{C1}	R_1
9	675	552.27	306.8	75
9	720	570.53	321.4	80
9	765	588.22	335.67	85
9	810	605.39	349.63	90
9	855	622.09	363.31	95
9	900	638.36	376.71	100
9	1125	714.14	440.24	125
9	1350	782.62	498.94	150
9	1575	845.58	553.81	175
9	1800	904.16	605.54	200
9	2025	959.17	654.64	225
9	2250	1011.19	701.48	250
9	2475	1060.66	746.36	275
9	2700	1107.93	789.51	300
10	10	50.5	9.17	1
10	20	87.18	17.2	2
10	30	112.47	24.74	3
10	40	133.04	31.91	4
10	50	150.83	38.8	5
10	60	166.73	45.45	6
10	70	181.25	51.89	7
10	80	194.68	58.16	8
10	90	207.24	64.26	9
10	100	219.09	70.23	10
10	110	230.33	76.06	11
10	120	241.04	81.78	12
10	130	251.3	87.38	13
10	140	261.15	92.89	14
10	150	270.65	98.29	15
10	160	279.82	103.61	16
10	170	288.7	108.85	17
10	180	297.32	114.01	18
10	190	305.7	119.09	19
10	200	313.85	124.1	20
10	210	321.79	129.05	21
10	220	329.55	133.93	22
10	230	337.12	138.75	23
10	240	344.53	143.51	24
10	250	351.78	148.22	25
10	260	358.89	152.87	26
10	270	365.86	157.47	27
10	280	372.69	162.03	28
10	290	379.41	166.53	29
10	300	386.01	170.99	30
10	320	398.87	179.78	32
10	340	411.34	188.4	34
10	360	423.44	196.87	36
10	380	435.2	205.2	38
10	400	446.65	213.38	40
10	420	457.82	221.44	42
10	440	468.72	229.37	44
10	460	479.37	237.19	46
10	480	489.8	244.9	48
10	500	500	252.5	50
10	550	524.64	271.07	55
10	600	548.18	289.07	60
10	650	570.75	306.56	65
10	700	592.45	323.58	70
10	750	613.39	340.18	75
10	800	633.64	356.37	80
10	850	653.26	372.21	85
10	900	672.31	387.7	90
10	950	690.83	402.87	95
10	1000	708.87	417.74	100
10	1250	792.94	488.23	125
10	1500	868.91	553.36	150
10	1750	938.75	614.25	175
10	2000	1003.74	671.66	200
10	2250	1064.78	726.14	225
10	2500	1122.5	778.12	250
10	2750	1177.39	827.92	275
10	3000	1229.84	875.8	300

low impedances sometimes encountered. Also, some measuring procedures require open- or short-circuits at either the input or output of the transistor under test. This is not always feasible at RF; not only is there the practical difficulty of producing a good short- or open-circuit, but such a procedure may provoke the transistor into oscillation, thereby ruining the measurement (and perhaps the transistor, too).

Being aware of this does not, in itself, alleviate confusion. However, it should also be kept in mind that the various impedance presentations can all be converted into one another's format. Often the conversion involves simple algebraic manipulation. And where the conversion process borders on the complex, the desired read-out is readily obtained from the scales on a Smith chart. In all instances, the objective is to derive either series- or parallel-equivalent impedance values so that the networks of Figure 4-7 can be solved. (Networks A, C_2, and D are shown with transistor impedance formats best handled by series-equivalent impedances. The other networks are more amenable to calculation via parallel-equivalent impedances with regard to the device to be matched. In any event, the two equivalent impedance formats are convertible into one another by means of the first four equations listed in Table 4-1.

In Figures 4-8 through 4-13 examples of various impedance formats are shown for randomly selected devices. The series- and parallel-equivalent impedances are illustrated, respectively, in Figures 4-8 and 4-9. Note that in one case capacitive reactance is plotted, whereas in the other ca-

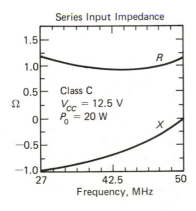 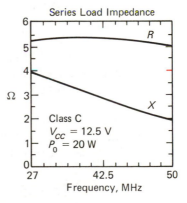

FIGURE 4-8

Series-Equivalent Impedance Information.

The data are readily convertible to parallel-equivalent information should the need arise. In certain networks, the series form facilitates computation of the network elements, however.

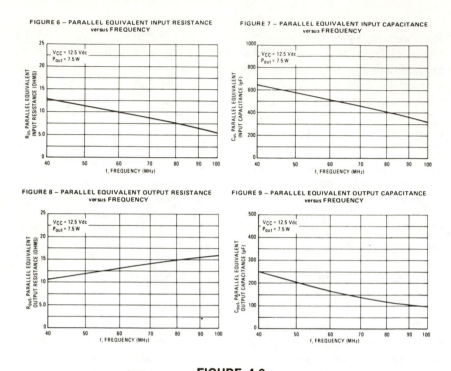

FIGURE 6 – PARALLEL EQUIVALENT INPUT RESISTANCE
versus FREQUENCY

FIGURE 7 – PARALLEL EQUIVALENT INPUT CAPACITANCE
versus FREQUENCY

FIGURE 8 – PARALLEL EQUIVALENT OUTPUT RESISTANCE
versus FREQUENCY

FIGURE 9 – PARALLEL EQUIVALENT OUTPUT CAPACITANCE
versus FREQUENCY

————————————— **FIGURE 4-9** —————————————
Parallel-Equivalent Resistances and Capacitances.
These examples of impedance information are suggestive of
physical models of transistor input and output circuits at low to
medium-high frequencies.

pacitance, itself, is shown. Either type of plot could be used with either
impedance format. We merely use the relationships $X_C = 1/2\pi f(C)$ or
$C = 1/2\pi f(X_C)$ to transform from capacitive reactance to capacitance,
and vice versa.

The constituents, conductance and susceptance, of admittance are
shown in the graphs of Figure 4-10. These quantities are related to the
constituents of impedance, resistance, and reactance, as follows:

$$R = \frac{G}{G^2 + B^2} \quad \text{and} \quad X = \frac{B}{G^2 + B^2}$$

where R = resistance in ohms
$\quad X$ = reactance in ohms (minus for capacitive; plus for inductive)
$\quad G$ = conductance in siemens
$\quad B$ = susceptance in siemens (plus for capacitive; minus for inductive)

Thus, the constituents of impedance, R and X, are readily calculated from the constituents of admittance, conductance and susceptance. Although conductance corresponds to resistance and susceptance corresponds to reactance, it is not *generally* true that conductance is $1/R$. Such a simple reciprocal relationship exists only when the reactance is zero. Likewise, a simple reciprocal relationship between susceptance and reactance exists only when the resistance is zero. These facts can be seen from the conversion equations

$$G = \frac{R}{R^2 + X^2} \quad \text{and} \quad B = \frac{X}{R^2 + X^2}$$

Whereas impedance, Z, obtains from $Z = \sqrt{R^2 + X^2}$, otherwise notated as $Z = R \pm jX$, admittance, Y, obtains from $Y = \sqrt{G^2 + B^2}$, other-

y PARAMETERS versus FREQUENCY

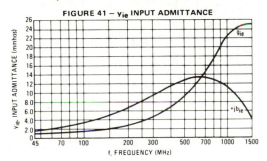

FIGURE 41 – y_{ie} INPUT ADMITTANCE

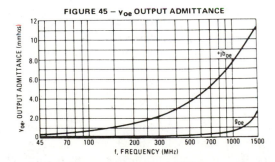

FIGURE 45 – y_{oe} OUTPUT ADMITTANCE

_____**FIGURE 4-10**_____
Transistor Input and Output Information
Presented in Terms of Admittance Data.
In some networks, admittance data are more relevant than the corresponding impedances.

SERIES EQUIVALENT IMPEDANCE

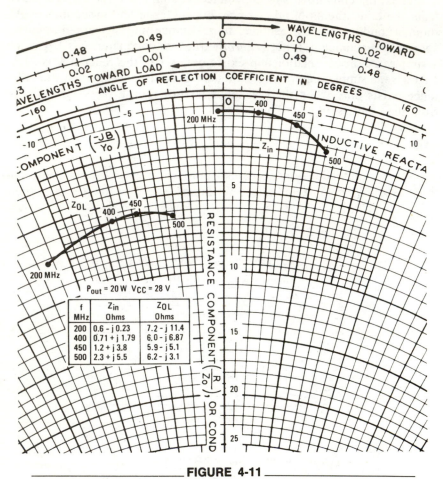

The table shown within the chart:

f MHz	Z_{in} Ohms	Z_{OL} Ohms
200	0.6 - j 0.23	7.2 - j 11.4
400	0.71 + j 1.79	6.0 - j 6.87
450	1.2 + j 3.8	5.9 - j 5.1
500	2.3 + j 5.5	6.2 - j 3.1

$P_{out} = 20\,W$ $V_{CC} = 28\,V$

_____ **FIGURE 4-11** _____
Smith Chart Display of Transistor Input and Output Impedance.
This vendor also presents the plotted measurements in tabular
form.

wise notated as $Y = G \pm jB$. Admittance and impedance are connected
by a simple reciprocal relationship, however. Thus, $Y = 1/Z$ and $Z = 1/Y$.

A common pitfall in the use of these entities is the fact that inductive
susceptance is negative, whereas inductive reactance is positive. And capacitive susceptance is positive, whereas capacitive reactance is negative.

Admittance quantities are useful in dealing with parallel circuits because the constituents, conductance and susceptance, combine arithmeti-

cally. Impedance quantities are useful in dealing with series circuits because the constituents, resistance and reactance, combine arithmetically. One of the beauties of the Smith chart is that both impedance and admittance quantities can be dealt with in the scaling system. It is only necessary that we be consistent when dealing in one or the other format.

A typical example of a Smith chart presentation of series-equivalent impedances is shown in Figure 4-11. It is easy to read out the resistive and reactive components for any frequency within the 200- to 500-MHz range. To facilitate reading the chart, this vendor also includes a tabular listing of the plotted resistances and reactances. Without this table, one could easily become confused in an effort to "normalize" the readings, as is done when using the Smith chart for transmission-line determinations. In this case, we need not concern ourselves with multiplying or dividing by 50, or other impedance values. The constituents of series impedance are simply read "as is."

Another way of dealing with the chart is to consider it normalized to 1 ohm. Then the notation R/Z_0 simply becomes R. And the $\pm jX/Z_0$ notations (which are not seen on the limited segment of the chart shown) simply become $\pm jX$. This is tantamount to saying that the chart is read "as

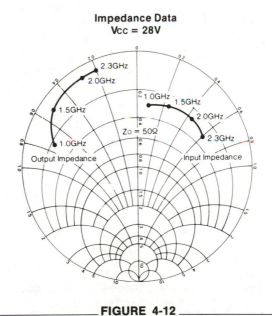

Impedance Data
Vcc = 28V

_____ **FIGURE 4-12** _____
Simplified Smith Chart Impedance Information.
All nonrelevant material has been deleted to facilitate read-out of
input and output impedances.

is.'' This viewpoint confirms the instructions delineated in the previous paragraph and may be a more comfortable approach for those accustomed to using the Smith chart for transmission-line problems.

Another vendor employs a graphically simplified Smith chart to display input and output impedances. Such a presentation is shown in Figure 4-12. It can be seen that all nonrelevant scales are deleted. It should be noted, however, that the chart is normalized to 50 ohms. Thus, all numerical values of resistance and reactance read from the chart must be multiplied by 50.

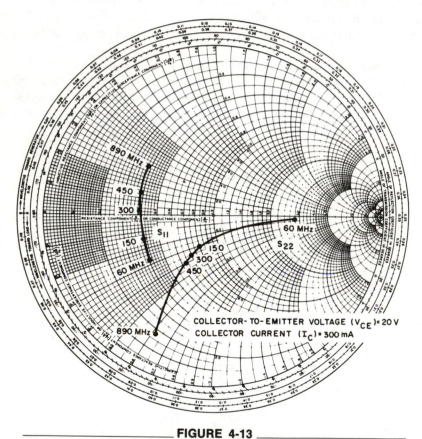

COLLECTOR- TO- EMITTER VOLTAGE (V$_{CE}$)= 20 V
COLLECTOR CURRENT (I$_C$) = 300 mA

_____ **FIGURE 4-13** _____
Indirect Impedance Display on Smith Chart
From Scattering Parameters.
Even though the manufacturer made the measurements in terms of scattering parameters, additional computations enable us to derive the corresponding input and output impedances.

● Scattering or s Parameters

Indirect impedance information obtained from scattering parameter measuring techniques is shown in the example of Figure 4-13. A subsequent mathematical or graphical procedure is required to convert such a presentation to actual input and output impedance data. It is only natural to ponder the reason for displaying the impedance information in such indirect fashion. It turns out that the manufacturer is able to make scattering parameter measurements more easily and more accurately than actual resistance or reactance measurements. When direct measurements of transistor input and output characteristics are made, it is necessary to use short- and open-circuits during the measurement procedure. However, such zero and infinite impedances are not easy to come by at radio frequencies, especially at VHF, UHF, and, certainly, microwaves. Even if they can be satisfactorily approximated, the transistor is often provoked into oscillation, thereby ruining the test (and often the transistor under test).

On the other hand, when scattering parameter techniques are used, the measured quantities are voltage ratios and reflection angles. It is somewhat like evaluating the performance of a transmitter-antenna system by observing the voltage standing-wave ratio. In this measurement process, the reflection coefficient is an indirect indication of impedance and can later be so converted. The significant aspect of this type of measurement is that the transistor under test is "enclosed" within 50-ohm impedances. Under this condition, oscillation or other instabilities are very unlikely, and the difficulties of approaching zero or infinite impedances are completely circumvented. Moreover, by means of 50-ohm coaxial cable, the manufacturer can locate measuring instruments remotely from the transistor undergoing evaluation.

The apparent disadvantage of this measurement technique is that the circuit designer must incorporate extra steps in transforming the *s* parameter data to impedance information. However, if he or she already has some proficiency with Smith chart procedures, the additional work is not tedious because it can be carried out by simple graphical constructions on the Smith chart. Indeed, the designer tends to adopt the language of reflection coefficients in dealing with the input and output characteristics of transistors. And, since high-frequency network elements are often transmission lines, or portions thereof, the *s* parameter display fits into the scheme of things.

However, it is not even necessary to make such graphical conversions on a Smith chart. Rather, the conversion formulas of Table 4-3 can be used. To be sure, these relationships, though straightforward, are a bit unwieldly because it is intended that the *s* parameters be expressed as magnitudes accompanied by phase angles; it is similarly intended that im-

——————————— TABLE 4-3 ———————————

Conversions Between s and Z Parameters

At the higher frequencies, s parameter measurement and evaluation
tends to be more practical than measuring impedances
or admittances.

$$s_{11} = \frac{(Z_{11} - 1)(Z_{22} + 1) - Z_{12}Z_{21}}{(Z_{11} + 1)(Z_{22} + 1) - Z_{12}Z_{21}} \qquad Z_{11} = \frac{(1 + s_{11})(1 - s_{22}) + s_{12}s_{21}}{(1 - s_{11})(1 - s_{22}) - s_{12}s_{21}}$$

$$s_{12} = \frac{2Z_{12}}{(Z_{11} + 1)(Z_{22} + 1) - Z_{12}Z_{21}} \qquad Z_{12} = \frac{2s_{12}}{(1 - s_{11})(1 - s_{22}) - s_{12}s_{21}}$$

$$s_{21} = \frac{2Z_{21}}{(Z_{11} + 1)(Z_{22} + 1) - Z_{12}Z_{21}} \qquad Z_{21} = \frac{2s_{21}}{(1 - s_{11})(1 - s_{22}) - s_{12}s_{21}}$$

$$s_{22} = \frac{(Z_{11} + 1)(Z_{22} - 1) - Z_{12}Z_{21}}{(Z_{11} + 1)(Z_{22} + 1) - Z_{12}Z_{21}} \qquad Z_{22} = \frac{(1 + s_{22})(1 - s_{11}) + s_{12}s_{21}}{(1 - s_{11})(1 - s_{22}) - s_{12}s_{21}}$$

pedance parameters incorporate magnitude and phase information. (In either instance, the parameters may, of course, be expressed in their rect-
angular-coordinate forms. For example, impedances may be expressed
either as magnitude at a phase angle, or as $R \pm jX$. Such notations stem
from complex algebra as commonly used in alternating-current circuits
and are not in any way peculiar to the concept of s parameters.)

In practice, s parameter information is readily obtained with the use
of vector voltmeters. These instruments directly provide the magnitude
and phase-angle information in their read-out format.

It should be borne in mind that the s parameters are reflection coefficients. Specifically, the following applies:

- s_{11} is the input reflection coefficient.
- s_{22} is the output reflection coefficient
- s_{21} is a reflection coefficient ratio that corresponds to forward
  transmission gain.
- s_{12} is a reflection coefficient ratio that corresponds to reverse
  transmission gain.

Those familiar with the use of impedance (Z) and admittance (Y) parameters will note the correspondences in the use of the numerical subscripts.

It may be of interest to contemplate that the widely used SWR meter
in amateur and CB transmitters indicates voltage standing-wave ratio only
because of scale calibration. In actuality, such meters respond to the ratio
of reflected to incident voltage from the load (antenna). Thus, the quantity

really being monitored is a reflection coefficient or *s* parameter. Indeed, the meter calibrations derive from the relationship

$$\text{SWR} = \frac{1 + s}{1 - s}$$

(No subscripts are used with *s* here to avoid confusion with the actual topic discussed, the application of *s* parameters to active devices such as transistors.)

In addition to the relationship of *s* parameters to SWR measurements, there also exist simple relationships to power quantities and ratios. These are often very useful in engineering design, as well as in operational appraisal. Such power relationships are listed in Table 4-4. As will be readily seen, each obtains from the square of an *s* parameter.

● **Impedance Matching: Circuits and Semantics**

The basic objective of impedance matching is properly explained in engineering texts. Therein, it is explained that the maximum transfer of power to a load takes place when the load has the conjugate impedance of the circuitry which drives it. For example, if a source has an internal impedance consisting of 10 ohms of resistance and 5 ohms of capacitive reactance, the load should have an impedance of 10 ohms of resistance and 5 ohms of inductive reactance. We are not surprised that this is so, for the process is clearly one of resonance, such that the net reactance of the overall circuit is canceled.

Unfortunately, many of these texts do not go beyond this idyllic, but not universally encountered, situation. For the preceding definition to be

_____ **TABLE 4-4** _____

By simply squaring the respective *s* parameters, additional useful performance criteria are provided.

$|s_{11}|^2 = \dfrac{\text{power reflected from the network input}}{\text{power incident on the network input}}$

$|s_{22}|^2 = \dfrac{\text{power reflected from the network output}}{\text{power incident on the network output}}$

$|s_{21}|^2 = \dfrac{\text{power delivered to a } Z_0 \text{ load}}{\text{power available from } Z_0 \text{ source}}$

$\quad\quad = $ transducer power gain with Z_0 load and source

$|s_{12}|^2 = $ reverse transducer power gain with Z_0 load and source

valid, one must assume that the source behaves essentially like a constant-voltage generator. The usual assumption is that the source *is* a perfect constant-voltage generator somewhat degraded from appearing as such because of internal resistance. The important point here is that this internal resistance is assumed to be relatively *low*.

The collector output circuit of a transistor does *not* comply with this assumption, however. Far from approximating a constant-voltage generator, it is a very good *constant-current* generator. This implies an extremely high internal resistance (see Figure 4-14). It would not be practical to attempt to match to this high internal resistance. Moreover, transistor parameters are considerably different in *large-signal* applications (class C, B, and AB) than in class A.

The manufacturer rates the transistor to operate under specified peak dc voltage and power dissipation conditions. By making an educated guess, we can postulate an expected output power from these data. Then an Ohm's law relationship can be used to calculate an apparent internal resistance of the transistor. This is approximately given by

$$R_L = \frac{V^2}{2(P)}$$

where R_L is the apparent internal resistance, V is the dc operating voltage, and P is the estimated power which will be delivered to the load. Then "impedance matching" is implemented between the output load imped-

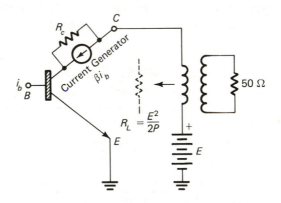

_____FIGURE 4-14_____
**Simplified Circuit Showing Output
"Impedance-Matching" Relationship.**
No attempt is made to match the high R_c resistance. Rather, the
50-ohm load resistance is transformed to the value R_L. In so doing,
P is the estimated amount of power delivered to the 50-ohm load.

ance and this apparent transistor output impedance. Power transfer under these conditions is optimum for the conditions under which the transistor is rated, but is not mathematically a maximum. Figure 4-14 shows the basic concept. For example, a transistor operating from a 30-V dc supply and intended to provide 5 W of load power would be postulated to have an apparent internal resistance of $30^2/(2 \times 5)$ or 90 ohms. A transformer or matching network would then be designed to transform this value to the 50-ohm load. At the same time, conjugate reactance would be incorporated to cancel whatever reactance was "seen" at the collector of the transistor.

For the purist, this impedance-matching procedure obviously deviates from the classical definition. In actual practice, it need be no stumbling block, however. Both matching procedures yield optimum load power under their respective circuit and rating conditions.

● A Closer Look at the Output Matching Situation

What has been discussed relating to "matching" the real or resistive part of the transistor's output impedance serves the purpose of achieving an approximation of proper performance. Among the simplifying assumptions made are the following:

- $V_{CE(sat)}$ is zero.
- There is no voltage drop in the dc collector feed system.
- The loaded Q of the output network is sufficient to produce a pure sine wave.
- All harmonics of the operating frequency "see" zero collector load impedance ($R_L = 0$).

None of these assumptions holds true in practical circuits. Various procedures can be resorted to in order to attain a more desirable match between the transistor and the load. For example, a more refined equation for the determination of R_L is

$$R_L = \frac{(E - V_{CE(sat)})^2}{2P}$$

where $V_{CE(sat)}$ can be obtained from specifications sheets as a function of collector current. It will often be in the vicinity of 1.5 to 2.5 V.

Aside from these assumptions, there are *other* reasons why experimentation is generally desirable in order to arrive at the best output network values. There is an interplay between matching conditions for optimum power output, power gain, and transistor operating efficiency. It is generally wise to sacrifice some power gain if, in so doing, higher operating efficiency can be realized. This need not necessarily require compro-

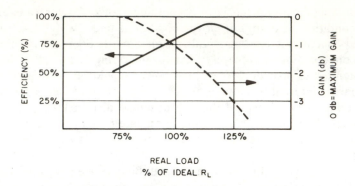

FIGURE 4-15
**Contradictory Optimization Requirements
for Efficiency and Power Gain.**
The ideal R_L is a compromise value which trades off a small
amount of power gain for an appreciable improvement in
efficiency. The effect on actual power delivered to the load is not
shown because of its additional dependencies on drive and
thermal conditions. Generally, the desired power can be realized
over a reasonable range of R_L values.
(Courtesy of Communications Transistor Corp.)

mise in the power output. Of course much depends upon the selection of the transistor, available drive, and the voltage regulation of the dc power source. Figure 4-15 illustrates the antagonistic relationship generally encountered between operating efficiency and power gain. Obviously, a relatively large improvement in efficiency can be purchased for a small reduction in power gain by making an appropriate change in the value of R_L.

Another complicating factor is that the transistor generally should not be hard driven into saturation. Both efficiency and reliability are well served if the transistor is operated at about 80% of its maximum available power. This harks back to the all-important initial selection of the device.

The value of R_L, together with the effective parallel capacitance of the collector, comprise the data for calculating the parallel equivalent output impedance of the transistor. In many cases, this capacitance is indicated in the specification sheet. It may not always be labeled clearly, but the general nomenclature "output capacitance," C_{OUT}, or similar description can be taken to indicate the *parallel* capacitance acting from collector to RF ground. It is qualitatively the same capacitance listed as C_{CB} for small-signal operation. Quantitatively, it is larger than C_{CB}, tending to be about twice as great. In any event, the parallel-equivalent output impedance of an RF power transistor is readily computed by forming a parallel circuit of this capacitance and R_L as determined from the operating voltage and power output.

For example, a transistor amplifier delivering 5 W to the load and operating from a 30-V supply computed to have an R_L value of 90 ohms. Suppose that it is found in the specification sheet that this transistor has an output capacitance of 100 pF at, or in the vicinity of, the frequency we wish to operate. Then the parallel-equivalent output impedance of the transistor corresponds to the impedance of 90 ohms of resistance in parallel with the reactance of the 100-pF capacitance. If the frequency of interest were 10 MHz, the reactance of the capacitance would be $1/2\pi fC$ ohms, or 159 ohms. This situation is depicted in (a) of Figure 4-16.

It is often more convenient to deal with the series-equivalent circuit.

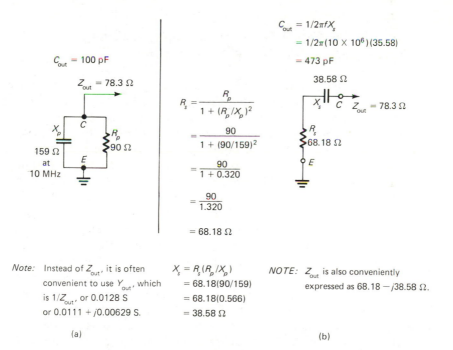

(a)

$C_{out} = 100\ pF$

$Z_{out} = 78.3\ \Omega$

159 Ω at 10 MHz

$R_s = \dfrac{R_p}{1 + (R_p/X_p)^2}$

$= \dfrac{90}{1 + (90/159)^2}$

$= \dfrac{90}{1 + 0.320}$

$= \dfrac{90}{1.320}$

$= 68.18\ \Omega$

Note: Instead of Z_{out}, it is often convenient to use Y_{out}, which is $1/Z_{out}$, or 0.0128 S or $0.0111 + j0.00629$ S.

$X_s = R_s(R_p/X_p)$
$= 68.18(90/159)$
$= 68.18(0.566)$
$= 38.58\ \Omega$

(b)

$C_{out} = 1/2\pi fX_s$

$= 1/2\pi(10 \times 10^6)(35.58)$

$= 473\ pF$

38.58 Ω

$Z_{out} = 78.3\ \Omega$

R_s
68.18 Ω

NOTE: Z_{out} is also conveniently expressed as $68.18 - j38.58\ \Omega$.

FIGURE 4-16

Equivalency between Parallel and Series Impedance Formats.
(a) Output impedance of transistor shown as a parallel *RC* circuit.
(b) Output impedance of same transistor shown as a series *RC*
circuit. The impedance of circuit (a) is $1/\sqrt{(1/R_p)^2 + (1/X_p)^2} = 78.3$
ohms. The impedance of circuit (b) is $\sqrt{(R_s)^2 + (X_s)^2} = 78.3$ ohms.

To change the parallel-equivalent output impedance to the series-equivalent output impedance, and vice versa, the conversion equations shown in Table 4-1 are used. Note that it is quite simple to convert from one circuit form to the other. Inasmuch as both output impedance formats will be encountered in the technical literature, such conversions tend to be a routine process when designing networks. The practicability of network element values and the ease of using the Smith chart will often dictate which circuit format is best for a given situation. Representation of the series-equivalent output impedance of the 5-W transistor amplifier is shown in Figure 4-16(b). It will be seen that some of the output networks in Table 4-2 utilize the parallel-equivalent circuit, while others make use of the series-equivalent circuit. As the term "equivalent" denotes, these circuit formats can represent the same transistor operating under like voltage and power conditions.

● What to Do About the Inductance of Coupling, Bypassing, and dc Blocking Capacitors

At low RF frequencies, say in the several megahertz region, the selection of capacitors for coupling, bypassing, and dc blocking is often made in a manner suggestive of audio-frequency practice: the capacitor is decreed to be "at least large enough" and is generally chosen to be several or ten times larger than this. The philosophy here is that the capacitor can hardly be too large from electrical considerations inasmuch as it should ideally behave as an ac short-circuit. Because of the parasitic inductance inherent in all capacitors, such a design motif is bound to produce less than optimum results at higher frequencies. Indeed, the oversized capacitor can behave as a high rather than a low impedance such as one would infer from naively using the capacitive reactance equation, $X_C = 1/2\pi fC$.

Whereas we were interested in simulating the reactance of an ideal network capacitor, the approach is somewhat different for coupling, bypass, and dc blocking capacitors. In these functions, the objective is to make the overall reactance as *low* as possible, or ideally, to eliminate it altogether. To accomplish this, the capacitor is treated as a series-resonant circuit. Every practical capacitor has its natural resonant frequency because of its parasitic or lead inductance. Except for varying the lead length, we do not have control of the inductive component of the capacitor. Lead-length adjustment suffices for relatively low frequencies, but generally it is desirable to use "leadless" capacitors. This is because series resonance with high inductance and relatively low capacity tends to produce high Q. This, in turn, makes the amplifier circuit more critical in adjustment and operation, and restricts bandwidth. Chip capacitors and uncased mica types are generally suitable for such self-resonant operation. These matters are illustrated in Figure 4-17.

As a consequence of the preceding considerations, the practical design procedure is to select that capacity value which resonates with its parasitic inductance at the operating frequency or at midband of the desired frequency range. Whenever possible, this resonance should be confirmed by some measurement technique or empirical test. For example, one can short the capacitor and couple a frequency dip meter to the short-

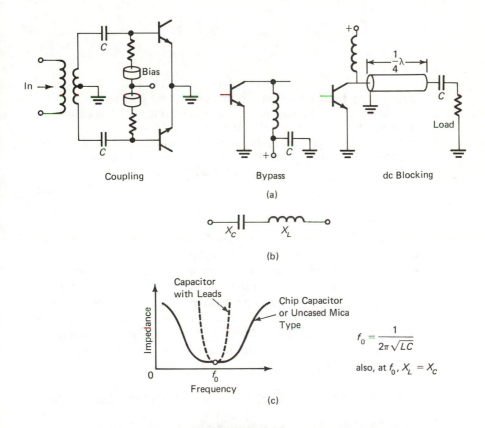

$$f_0 = \frac{1}{2\pi\sqrt{LC}}$$

also, at f_0, $X_L = X_C$

FIGURE 4-17

Design Factors Pertaining to Coupling, Bypass, and dc Blocking Capacitors.
The basic objective is to use self-resonance to attain low impedance. (a) Circuit positions of coupling, bypass, and dc blocking capacitors. (b) Equivalent circuit of actual capacitors, showing parasitic or lead inductance. (c) Series resonance mode used for producing low impedance.

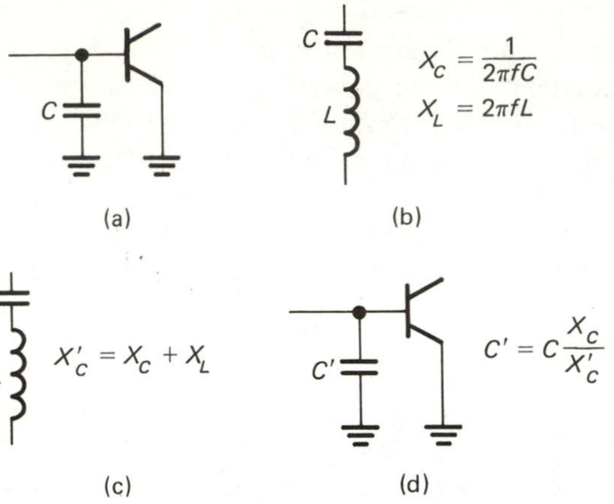

$$X_c = \frac{1}{2\pi fC}$$

$$X_L = 2\pi fL$$

(a)　　　　　(b)

$$X'_c = X_c + X_L$$

$$C' = C\frac{X_c}{X'_c}$$

(c)　　　　　(d)

FIGURE 4-18

Accounting for the Parasitic Inductance of RF Capacitors.
(a) Situation with ideal capacitor, C (no inductance). (b) The actual
electrical situation. L is the lead or parasitic inductance of
capacitor. (c) Recalculation of the capacitive reactance to simulate
ideal capacitor. (d) Implementation of new capacitor in circuit. C'
is measured at a low frequency.

ing conductor. Inasmuch as the short, itself, has inductance, a little skill
must be developed to obtain meaningful results.

The functional difference between coupling, bypass, and dc blocking
capacitors is not always sharply defined. Indeed, sometimes such capacitors are also used to help tune the impedance-matching networks. In such
instances, they would be treated as network capacitors, and the objective
would not be to have them self-resonate. However, their self-inductance
would still have to be accounted for in the manner to be subsequently described for network capacitors.

● What to Do About the Inductance of Network Capacitors

As we go up in frequency, the inductance in the leads and in the internal hardware of capacitors assumes ever greater importance. This is true
even of chip capacitors, uncased mica capacitors, and other special RF
types. Elsewhere, the admonition is given to be sure that the capacitor
used has the proper capacitance at the *actual operating frequency*. And it
is pointed out that the low-frequency capacitance is generally incorrect
for high-frequency use. Some of this deviation can be attributed to the
lump effects of measurement error, production tolerance, field fringing,

radiation, proximity influences, and frequency-dependent dielectric constant. However, the parasitic inductance of the capacitor often is the most important factor involved. One cannot naively assume RF capacitors to be ideal elements. Manufacturers of RF capacitors usually indicate the value of this internal inductance under such headings as "lead," "parasitic," "internal," or "residual" inductance. This being the case, one then specifies a *smaller* capacitor than that calculated on the basis of a "pure" capacitance. The idea is to have the *actual* series *LC* circuit simulate the behavior of the originally calculated ideal capacitor.

The procedure for accomplishing this is shown in Figure 4-18. Circuit (a) has the ideal capacitor, *C*. Figure 4-18(b) shows the equivalent circuit of a real-world capacitor with *L* representing the parasitic inductance. Inasmuch as series inductive reactance *subtracts* from capacitive reactance, we must *add* the value of the inductive reactance to the capacitive reactance to produce the *net* capacitive reactance actually needed. Thus, the new capacitor, having higher capacitive reactance than was calculated for the ideal capacitor, has *less* capacitance than the original ideal capacitor. In the implementation of this logic, it is assumed that the new capacitor has the same parasitic or lead inductance as the original one. This can be a valid assumption if one sticks to the same brand and type. Thus, in Figure 4-18(c), *L* is assumed to be the same value for new capacitor *C'* as for original capacitor *C*.

Suppose, for example, one calculates a capacitive reactance of 10 ohms to be used as a shunt element from base to ground. It is then found that the actual capacitor with 10 ohms of capacitive reactance *also* has 5 ohms of inductive reactance. The next procedure is to select a new capacitor with 15 ohms of capacitive reactance and of the same type as the originally calculated capacitor. We now have 15 ohms of capacitive reactance in series with 5 ohms of inductive reactance. This *combination* is the approximate equivalent of an ideal capacitor with 10 ohms of capacitive reactance. A penalty is paid in bandwidth, however, for our *LC* "equivalent" is frequency sensitive in the sense that it simulates the ideal capacitor only over a narrow frequency range.

Figure 4-18(d) uses the recalculated, or new, capacitor in the circuit (a). Inasmuch as the parasitic inductance is not generally shown in such schematics, it is not depicted in circuit (d). Because of its presence, however, capacitor *C'* of circuit (d) is smaller than the ideal capacitor (no inductance) of circuit (a). For simplicity, the transistor is not shown in (b) or (c). The value *C'* is its low-frequency, or "dc" value.

● Use of Two Parallel Capacitors

Although almost any amount of parasitic inductance can be accounted for by this technique, the frequency sensitivity of the *LC* combination becomes progressively worse with higher values of inductance. In the interest of broadbanding, it is common practice to parallel two like capacitors rather than use a single capacitor for C'. That is, two $C''/2$ capacitors are installed. This reduces the effective parasitic inductance by half, so we only need contend with $L/2$. (The use of such parallel capacitors not only makes the capacity simulation less frequency sensitive, but halves the RF current circulating in each capacitor. Also, a more balanced RF current path can be established in the PC board. To further enhance these features, more than two parallel capacitors are sometimes employed.)

It should be noted that the values of the two C'' capacitors are *not* simply one-half the value of the previously determined C' capacitor. This need not be surprising, for we have already seen how series inductance made it necessary to alter the ideal capacitor, C, to C'. If now we reduce the effective inductance by paralleling two capacitors, the equation $X_C' = X_C + X_L$ will no longer apply to our situation. Rather, a new capacitive reactance, X_C'' will be needed. And the capacitive reactance per branch will then be $2X_C''$. The equation modified to apply to our new situation is $X_C'' = X_C + \frac{1}{2}X_L$. (Note that L and X_L per capacitor are assumed the same whether we are dealing with C, C', C'', or as will now be the case $C''/2$.) This is because all these capacitors are assumed to have the same leads and very nearly the same construction features.

The logic and procedure for determining the values of the parallel capacitors are shown in Figure 4-19. The same reasoning and assumptions involved in Figure 4-18 apply here. Comparing step (c) in both procedures, we see that our net capacitance, C'', is *larger* than the single capacitor, C'. If our objective is to use parallel capacitors, there is no need to calculate C'. In any event, $C''/2$ represents the low-frequency capacitance of the individual capacitors.

● Harmonic Filters

Solid-state RF amplifiers are often broadbanded and therefore are capable of supplying considerable harmonic energy to the antenna system. Even when not appreciably broadbanded, the operating Q of the output network is generally quite low in order that the transistor can develop high efficiency. This, of course, is accomplished at the expense of harmonic discrimination. Some help is provided by linear, rather than class C operation, and push-pull circuitry materially reduces generation of even-order harmonics. Nonetheless, the practical implementation of most amplifiers requires consideration of harmonic repression ahead of the antenna. This calls for the insertion of a filter between the transmitter and the antenna feeder line. Such a filter must exhibit the following features:

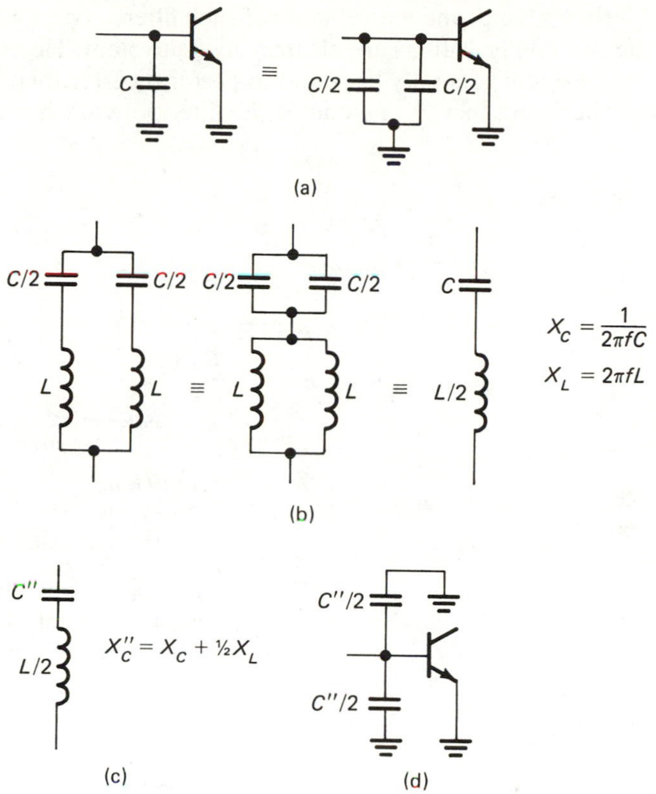

$$X_C = \frac{1}{2\pi fC}$$

$$X_L = 2\pi fL$$

$$X_C'' = X_C + \tfrac{1}{2}X_L$$

FIGURE 4-19

**Reducing Effective Parasitic Inductance
Via Paralleled Capacitors.**

(a) Situation with ideal capacitors. (b) The equivalent electrical situation of the parallel capacitor circuit. (c) Recalculation of net capacitive reactance to simulate ideal capacitor. (d) Implementation of new parallel capacitors in circuit. $C''/2$ is measured at a low frequency.

- Appreciable attenuation of harmonics must be provided.
- Minimal insertion loss at the fundamental frequency is essential.
- The filter must not cause excessive degradation of the VSWR.

All things considered, simple pi and tee networks designed by easily handled image-parameter principles generally prove satisfactory. Two cascaded low-pass, full sections provide a readily implemented network which will satisfactorily filter out harmonic energy in most instances. Fig-

ure 4-20 shows the pi and tee versions of such filters. To a first approximation, the two configurations are electrically equivalent. However, in practice, one type will generally be found to perform better than the other. (It
may also be found that, if one end of the filter network has the shunt ca-

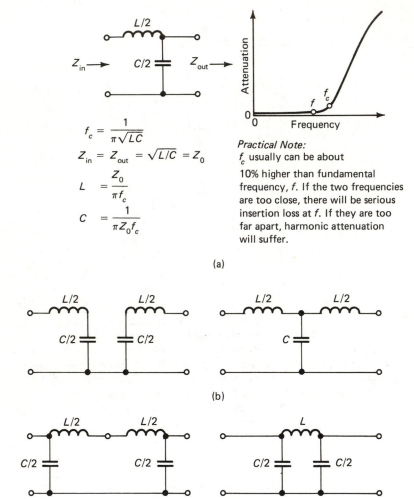

FIGURE 4-20
Image Parameter Low-Pass Filter.
(a) The L section "building block" with basic formulas and
generalized response curve. (b) Synthesis of a full tee section. (c)
Synthesis of a full pi section.

pacitor and the other end has the series inductor, optimum results obtain. Which end of the filter connects to the transmitter must be determined empirically.)

A basic consideration attendant to the use of such low-pass filters is where the fundamental frequency will be allowed to fall on the response curve. The closer it is to the cutoff frequency, the more the harmonics will be attenuated. However, both insertion loss and VSWR will be adversely affected if the fundamental is too close to the cutoff frequency, which already is 3 dB down. Because of the tolerances, finite Q's, and parasitic reactances characterizing actual capacitors and inductors, some safety margin must be allowed in the design of the cutoff frequency. Generally, it is found that if the cutoff frequency is designed to be 10% higher than the fundamental, a reasonable compromise between conflicting factors is achieved.

Before attempting implementation of RF filters, the following considerations should be dealt with:

- The Q's of the inductors and capacitors should be high—in the vicinity of several hundred at least. This should appertain in the region of the cutoff frequency. Moreover, the Q's should be no lower than about 100 at the fifth harmonic of the fundamental.

- The tolerance of inductor and capacitor values should, if feasible, be known to ±5%. Tolerances greater than ±10% can readily lead to malperformance, although acceptable operation can usually be regained by "tweaking" techniques. Tolerances can be meaningless unless they apply to the general frequency range embraced by the cutoff frequency of the filter.

- Inductors and capacitors must be selected which have minimal parasitic reactance. This applies mostly to distributed capacitance in the inductor and to series inductance in the capacitors. In the latter case, this implies either very short leads or leadless constructions, such as capacitor "chips."

- The individual inductors in a filter must not be electromagnetically coupled (except in certain balanced configurations which are not commonly encountered). Inductors should not be too close to one another and should be oriented at right angles to each other.

- Care must be exercised to ensure that undesired coupling does not exist between the input and output of the filter. Such coupling can be inductive or capacitive, or can exist via radiation. Shielding techniques and the use of toroidal inductors are helpful.

● Example of Design of 30-MHz Harmonic Filter

The objective is to pass 30 MHz with minimal insertion loss and degradation of VSWR while imparting attenuation to the second, third, and higher harmonics. It is assumed that the filter is to be inserted into a nearly flat 50-ohm antenna feederline. The design procedure is as follows:

1. From the preceding paragraphs, it appears reasonable that a two-section pi low-pass configuration could be expected to fulfill the requirements. The characteristic impedance of such a filter will be the same as the transmission line (50 ohms). Such a filter is shown in Figure 4-21. Although more harmonic attenuation per filter element might be forthcoming from more sophisticated filter designs, this image-parameter type is exceptionally straightforward in computation and implementation.

2. Decide on the cutoff frequency, f_c. The rule of thumb here is to

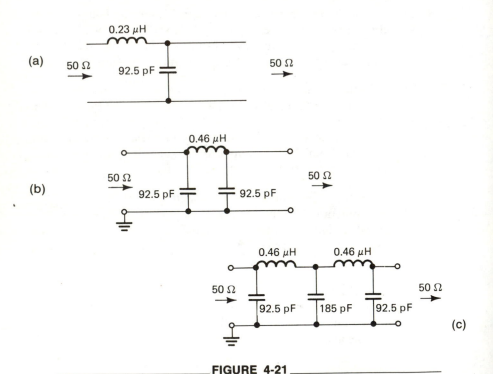

FIGURE 4-21

Synthesis of the Harmonic Filter for a 30-MHz Amplifier.
(a) Half-section building block with $Z_0 = 50$ ohms and $f_c = 34.5$ MHz. (b) Single pi section. (c) Two cascaded full-pi sections.

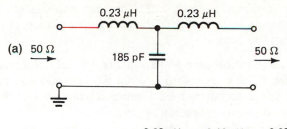

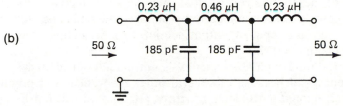

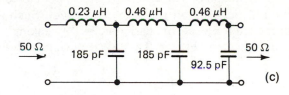

(c)

_____ **FIGURE 4-22** _____
Other Versions of the 30-MHz Harmonic Filter.
(a) Single tee section. (b) Two cascaded full tee sections. (c) Filter
with one tee termination and one pi termination.

make f_c about 10% or 15% greater than the fundamental frequency. Therefore, 115% × 30 MHz = 34.5 MHz = f_c.

3. Calculate the L and C values:

   a. $L = \dfrac{Z_0}{\pi f_c}$

   $= \dfrac{50}{(\pi)(34.5 \times 10^6)} = 0.461 \ \mu H$

   and $L/2 = 0.23 \ \mu H$.

   b. $C = 1/\pi Z_0 f_c$
       $= 1/(\pi)(50)(34.5 \times 10^6)$
       $= 0.000185 \ \mu F = 185 \ pF$
       and $C/2 = 92.5 \ pF$.

4. Put the $L/2$ and $C/2$ building blocks together to form the two full-

pi section filter as shown in Figure 4-21. Consideration should also be given to the single section of (b) if the harmonic products of the amplifier are not too severe.

Alternative filter configurations are shown in Figure 4-22. Note that the tee-pi filter of Figure 4-22(c) necessarily comprises two and one half sections. On this basis, alone, its harmonic attenuation will be better than either the two-section pi or the two section tee networks.

A study of these filters reveals that one cannot change a pi input or output section to a tee section merely by omitting the end capacitor. Nor can one change a tee input or output section to a pi section merely by omitting the end inductor.

Because the fundamental frequency is not made coincident with the cutoff frequency and because of several other theoretical and practical factors, the oft-quoted 6 dB/element/octave attenuation will not be realized. Four and one half to five decibels per element per octave is generally the practical roll-off rate one can expect. Such attenuation can often be exceeded for a specific harmonic by "tweaking" element values, but one should keep an eye on the VSWR during such experimentation.

5

Applications of Transmission-Line Elements to RF Power Circuitry

Semiconductor devices and network theory have been discussed and emphasis has been placed upon their *interdependency* in solid-state RF circuitry. Because of the ever-increasing utilization of VHF, UHF, and microwave spectral regions, the impedance-matching, resonating, and filtering networks are often comprised from transmission-line elements, rather than from "lumped" inductors and capacitors. Also, such elements often prove better performers than conventional RF chokes or even the best bypass capacitors.

Transmission-line theory is readily available in many good textbooks, but the treatment generally serves the purposes of the mathematician rather than the practical engineer, experimenter, or skilled hobbyist. This chapter endeavors to present this theory in such a way that practical implementations will obviously benefit.

Relevant, also, is the discussion of *transmission-line transformers*. These amazing devices represent a relatively new energy-transfer technique, one in which "primary" and "secondary" circuits are not coupled via electromagnetic induction in the manner of conventional transformers. Transmission-line transformers are destined to play very important roles in solid-state RF power. A good illustration of this is the kilowatt solid-state RF amplifier discussed in Chapter 7.

● Use of Transmission-Line Elements

Although "lumped circuit" reactances and transformers are much used in solid-state RF power systems, superior performance generally can be had with transmission-line elements. This is particularly true at VHF, UHF, and certainly in the microwave region, where stripline techniques contribute greatly to predictability, reproducibility, and low losses. Transmission-line devices can be used to simulate inductance, capacitance, series- and parallel-resonant circuits, and RF chokes. They also are widely employed for impedance transformation and for balun techniques (transforming from balanced to unbalanced circuitry, and vice versa). They are, additionally, much used in a variety of power splitters and power combiners. These function essentially as hybrid transformers and enable high power to be developed in the load as the summation of contributions from a multiplicity of amplifiers. At the same time, the individual amplifiers are electrically isolated from one another.

Transmission-line elements can take several physical forms. Parallel, coaxial, and twisted wire lines are commonly used. Stripline elements make use of single lines working against a ground plane. Such lines are popular because of their compatibility with printed-circuit-board techniques. Also, lines may be loaded in various ways with dielectric and ferromagnetic materials. At low frequencies, where transmission-line devices tend to be awkwardly long, they are often coiled up to achieve physical compactness. Coaxial cable is, in particular, amenable to this practice, and the electrical characteristics of such a coiled line generally remain unaltered for the purpose served.

The behavior of transmission lines is governed primarily by their electrical length and their characteristic impedance. The electrical length of practical lines is always somewhat shorter than their physical length. This is because the velocity of electromagnetic propagation along a line is less than it is in free space. Depending on the dielectric material, practical lines are often on the order of 60% to 98% of their length calculated on the basis of free-space velocity. The characteristic impedance of lines depends on constructional features and on dielectric or ferromagnetic materials inserted between the lines. Characteristic impedance values between 20 and 200 ohms are often encountered, but higher and lower levels can be attained where inordinately great or small spacing can be accommodated. In any event, the impedance "seen" by the circuit terminations will depend also upon the manner in which the lines are used. Schematic diagrams of RF circuits are often drawn with the L and C components that the transmission-line elements simulate.

● Characteristic Impedance of Transmission Lines

In the many different uses involving transmission lines in solid-state RF circuitry, the behavior of the line is invariably governed by its characteristic impedance and its effective electrical length. Characteristic impedance is a function of the line's construction and geometry and of the dielectric constant of the insulating medium. The effective electrical length depends primarily upon the dielectric constant, but may be also influenced by "proximity" effects from other conductors or insulators and by fringing or distortion of the electric lines of force in the insulating medium. Finally, if any magnetically permeable substance is interposed between the lines, both the characteristic impedance and the effective electrical length will be affected. Transmission lines in which air is the dielectric medium will, ideally, have very nearly the same physical and electrical lengths. In practice, other effects generally make the physical length slightly less than the effective electrical length. Only in "free space" and under ideal conditions are the physical and electrical lengths identical. In real life, the physical length will always be less than the effective electrical length, sometimes negligibly so, as in certain air-dielectric lines, sometimes considerably so, as in lines employing other dielectric substances. The physical length of a transmission line is $1/\sqrt{e}$ times its electrical length, where e is the dielectric constant of the substance between the line elements. For air, e is very close to its "free space" value of 1. All other commonly used substances have higher dielectric constants.

From the preceding, it is apparent that the parameters of a transmission line tend to be interrelated. Our insight into the nature of transmission-line applications to solid-state RF circuits can be enhanced by contemplating the following situations:

The characteristic impedance, Z_0, of any type or length of line can be determined by making two measurements, the input impedance of the line when the far end is open-circuited, and the input impedance of the line when the far end is short-circuited. The equation describing this relationship is, $Z_0 = \sqrt{Z_{oc} \times Z_{sc}}$, where Z_{oc} is the input impedance with the far end of the line open-circuited, and Z_{sc} is the input impedance with the far end short-circuited. This classical relationship is depicted in Figure 5-1.

Although this relationship is a measurement technique rather than an ordinary operating mode, the impedance value thereby obtained for Z_0 has a very special significance in transmission-line theory. A line of any length terminated at its far end by Z_0 will not reflect energy back to the source (input), but will absorb all energy reaching it. The only other situa-

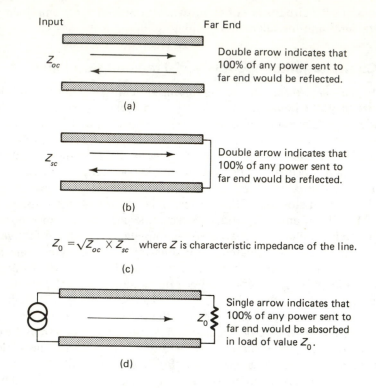

Input

Z_{oc}

Double arrow indicates that 100% of any power sent to far end would be reflected.

(a)

Z_{sc}

Double arrow indicates that 100% of any power sent to far end would be reflected.

(b)

$Z_0 = \sqrt{Z_{oc} \times Z_{sc}}$ where Z is characteristic impedance of the line.

(c)

Z_0

Single arrow indicates that 100% of any power sent to far end would be absorbed in load of value Z_0.

(d)

_____ FIGURE 5-1 _____
Characteristic Impedance of a Line from a Classical Measurement Technique.
(a) Z_{oc} is input impedance with far end open-circuited. (b) Z_{sc} is input impedance with far end short-circuited. (c) Z_0 may be calculated from Z_{oc} and Z_{sc}. (d) With Z_0 as far end termination (load), no power is reflected back to input.

tion where no reflections from the far end of the line back to the input take place is for an infinitely long transmission line. Thus, the Z_0 terminated line simulates the infinitely long line in this respect. In most RF applications, Z_0 is almost purely resistive, so it is often symbolized as R_0.

● **Quarter-Wave Lines**

Transmission lines which are electrically one-quarter of a wavelength display unique properties; these are readily put to good use in solid-state RF circuits. At the top of Figure 5-2 are shown unbalanced and balanced versions of the quarter-wave impedance transformer. Other physical real-izations of transmission lines, such as twisted wires or stripline elements, are equally applicable. The interpretation of the circuits is that a resistive termination of value $R1$ can be transformed to value $R2$ (or vice versa) providing that the characteristic impedance, Z_0, of the line is equal to $\sqrt{R1 \times R2}$. This relationship applies to both the unbalanced and bal-anced forms.

The center and bottom illustrations in Figure 5-2 show how the quar-ter-wave lines can be used to simulate parallel and series-resonant tanks. These arrangements can be used as the frequency-determining circuit in oscillators, as RF chokes, or as bypass elements.

To produce a large transformation between the terminating resist-ances of the impedance transformer, it may prove convenient to accom-

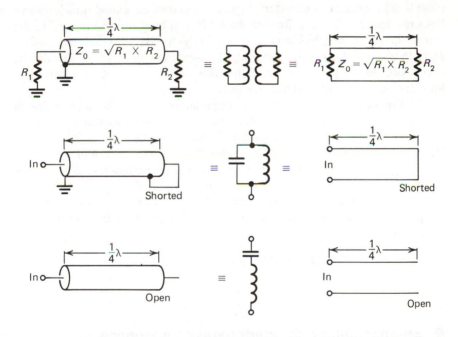

_____ FIGURE 5-2 _____

Unbalanced and Balanced Quarter-Wave Transmission-Line Functions.

Unbalanced circuits are shown on the left; balanced circuits are shown on the right. For the impedance transformer, the lines must have a certain characteristic impedance. For simulation of the LC resonant circuits, the value of the characteristic impedance is generally not critical.

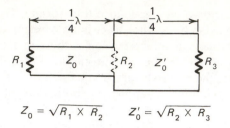

$$Z_0 = \sqrt{R_1 \times R_2} \qquad Z_0' = \sqrt{R_2 \times R_3}$$

_____ **FIGURE 5-3** _____

Cascaded Arrangement of Quarter-Wave Transformation Lines.
R2, a nonphysical resistance, is common to both quarter-wave
sections. Overall transformation ratio is higher than readily
attained with a single line.

plish this in more than one step. Figure 5-3 shows cascaded quarter-wave
lines used in this fashion. Resistance $R2$ is a "phantom resistance" in the
sense that it has no physical existence. Its value is, nonetheless, "seen"
by both quarter-wave lines. Additional cascading can be used. Besides its
use to provide large transformations, this technique is often resorted to
for developing broadbanded response.

In Figure 5-3 the characteristic impedance, Z_0', of the second line is
higher than that of the first line, Z_0. This is symbolized by the wider spacing of the elements in the second line. This being the case, it follows that
$R2$ is greater than $R1$, and $R3$ is greater than $R2$. Thus, the arrangement
transforms from a low resistance, $R1$, to a high resistance, $R3$, or vice
versa. However, $R1$ and $R3$ cannot be interchanged.

From a mathematical viewpoint, it is not necessary to deal with the
nonphysical resistance, $R2$. The formulas for the two line impedances
then become

$$Z_0 = \sqrt{R3 \times R1^3} \quad \text{and} \quad Z_0' = \sqrt{R1 \times R3^3}$$

● Broadbanding with Quarter-Wave Impedance Transformers

A useful feature of the cascaded quarter-wave impedance transformer
is its broadened frequency response. It turns out that this broadbanding
effect increases with the number of sections and is greater for lower overall ratios of impedance transformation. These characteristics are illustrated by the selectivity curves of Figure 5-4. With a little ingenuity, one

can implement simple, but effective, broadband impedance transformers at UHF and especially at microwave frequencies. The coiling of coaxial cable is permissible where problems are encountered with physical length. And microstrip transmission line need not necessarily be laid out as a straight run, but can incorporate bends to conserve linear space. Another useful aspect of the multiple-section impedance transformer is that its geometric and electrical parameters are less critical than single-section transformers in applications where broadbanding is not the first-priority operating feature.

● Quarter-Wave Lines Applied to Push-Pull Circuits

An almost too good to be true use of quarter-wave lines is shown in Figure 5-5. Here the lines perform simultaneously as baluns, impedance transformers, and phasing circuits for proper push-pull operation of the

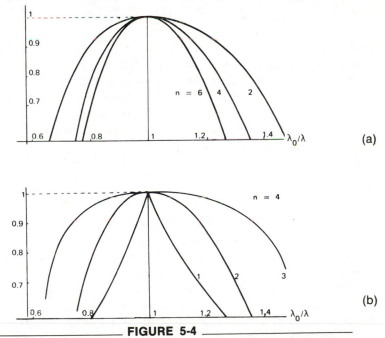

FIGURE 5-4

Selectivity Curves for Quarter-Wave Impedance Transforming Systems.

Broadbanding increases with more sections and with lower overall ratios of impedance transformation. (a) Two quarter-wave sections as shown in Figure 5-3. (b) One, two, and three quarter-wave sections; overall impedance transformation ratio, $n = 4$. (Courtesy of Motorola Semiconductor Products, Inc.)

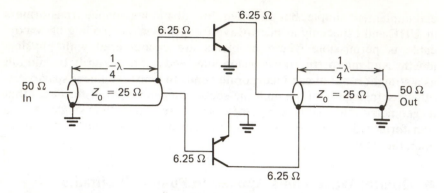

FIGURE 5-5
**Quarter-Wave Lines Used for Impedance Matching, Phasing, and
as Baluns in a Push-Pull Amplifier.**
System is exceptionally simple, but a fortuitous combination of
transistor operating conditions and line impedances is needed.

two transistors. The arrangement is depicted in simplified form in order to
focus attention on the implementation of the lines. Thus, various blocking
capacitors, RF chokes, and reactance-canceling components are omitted.
These, however, can be readily incorporated in the scheme in the same
way as in more conventional amplifier circuits.

In this example, the resistive components of both the input and output
impedance of the individual transistors are, fortuitously, 6.25 ohms.
Thus, both input and output lines "see" 12.5 ohms at their transistor terminations. The 25-ohm lines fit the situation beautifully, inasmuch as
$25 = \sqrt{12.5 \times 50}$. This takes care of the transformation feature of the
lines.

The lines comply with the push-pull aspect of the amplifier circuitry.
The base signals are 180° out of phase with one another. By reciprocal action, the 180° out-of-phase signals from the collectors are converged as a
single composite signal in the output line.

Finally, these quarter-wave lines function as baluns. The two base
signals are balanced with respect to ground. And the output signal delivered to the load is unbalanced to ground, even though the collector signals
are balanced to ground.

● **Quarter-Wave Lines in Transistor Circuits**

The quarter-wave line provides its usually desired transformer action
only when terminated in real (resistive) impedances. However, transistor
input and output impedances are generally complex: they have *both* resis-

tive *and* reactive components. To properly accommodate a quarter-wave line transformer, the reactive component must be tuned out. An example is shown in Figure 5-6(a) where a quarter-wave line is used to transform the resistive portion of the transistor input impedance to the desired 50 ohms. At the higher frequencies the input reactance of transistors tends to be inductive from the effect of the internal bonding wires. Therefore, an equal and opposite reactance is incorporated to cancel the transistor's input reactance. This takes the physical form of a capacitor or a section of transmission line selected to simulate the required capacitance. Capacitor *C* serves this purpose in Figure 5-6(a). This makes the quarter-wave line perform properly, and it then provides transformation between the *resistive* component of the transistor's input impedance and the desired 50 ohms.

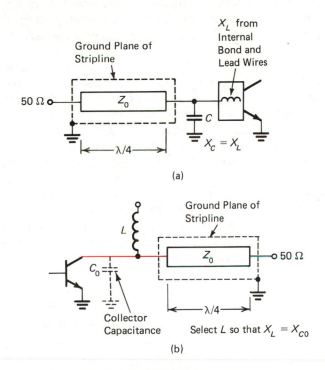

(a)

(b)

FIGURE 5-6

Use of Quarter-Wave Lines in the Input and Output Circuits of Transistors.

(a) Application to input of UHF or microwave transistor. (b) Application to output circuit of transistor. In both cases, $\lambda/4$ is the electrical length of the line and $Z_0 = \sqrt{R1 \times R2}$, where $R1$ and $R2$ are the terminating resistances "seen" by the line.

Figure 5-6(b) shows the same principle applied to the output of a transistor amplifier. Here the undesired reactance is capacitive, so an *inductive* reactance is needed. This is often accomplished by a small lumped circuit inductor. From its schematic position, this inductor appears to function as an RF choke. However, instead of complying with the simple requisite of providing high inductive reactance, the stipulation here is for a discrete value of inductive reactance. Of course, such an inductor also can serve as a dc feed path. In practice, it is common to use both this resonating inductance and an RF choke, together with the usual decoupling circuitry. In any event, the quarter-wave line then performs its transformation between *resistive* terminations.

● Eighth-Wave Line

Like the quarter-wave transmission line, one having an electrical length of one-eighth wavelength displays some uniquely useful circuit properties. And, similarly, the use of these properties affords the designer and experimenter interesting ways to match impedances in solid-state circuits. A peculiarity of the eighth-wave line is that its input end always "looks" resistive when its opposite end is terminated in the magnitude of the line's characteristic impedance. Note that this statement allows complete latitude in how this impedance magnitude is comprised; it can be made up of any combination of resistance and/or reactance that produces an impedance magnitude equal to the characteristic impedance of the line, Z_0. That is, the input to such a line will appear as a pure resistance if the terminated end "sees" the impedance $Z_0 = \sqrt{R^2 + X^2}$. Here Z_0 is the characteristic impedance of the line specified as a simple magnitude, such as 50 or 72 ohms. R and X are the termination resistance and reactance, respectively. (The termination must be expressed as a series-equivalent circuit. If parallel-equivalent values are available for transistor impedances, they must first be converted to the series-equivalent format. (See Table 4-1.)

The fact that a resistive input can be achieved despite the presence of reactance at the terminated end is, indeed, interesting. But what about the *value* of the input resistance? It is found from the equation

$$R_{\text{IN}} = \frac{R}{1 - \dfrac{X}{\sqrt{R^2 + X^2}}}$$

where R is the resistive component of the terminating impedance and X is the reactive component of the terminating impedance. Figure 5-7 shows an example in which the base of a transistor is the terminating impedance.

It is always possible to implement an eighth-wavelength line to provide a purely resistive input. Microstrip lines enable considerable flexibil-

$$20 + j15 \ \Omega \equiv \sqrt{(20)^2 + (15)^2} = 25 \ \Omega$$

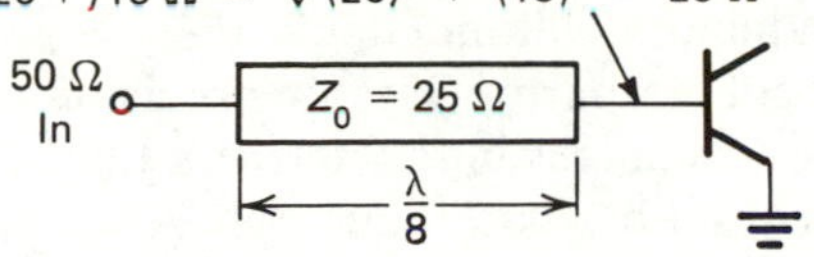

$$R_{in} = \frac{R}{1 - \dfrac{X}{\sqrt{R^2 + X^2}}} = \frac{20}{1 - \dfrac{15}{\sqrt{(20)^2 + (15)^2}}}$$

$$R_{in} = \frac{20}{1 - \dfrac{15}{25}} = \frac{20}{\dfrac{10}{25}} = \frac{500}{10}$$

$$R_{in} = 50 \ \Omega$$

FIGURE 5-7

Microstrip Eighth-Wavelength Line Used to Match Base Impedance to 50-Ohm Source Resistance.
Equal magnitudes of base impedance and the characteristic impedance of the line result in a purely resistive input. The *magnitude* of this input resistance is determined by the reactive and resistive components of the base impedance. "Matching" is produced in this example because the base impedance is transformed to the desired 50-ohm input resistance.

ity in this respect because they can be readily constructed to have characteristic impedances from several ohms to several hundred ohms. (It is not always easy to produce the 50-ohm input impedance, however. This often involves appropriate selection of transistors and operating conditions.) The eighth-wave matching-technique is applicable to the output of the transistor also. Note, however, in this example, that had the terminating impedance been capacitive, that is, $20 - j15$ ohms, the resistive input would have had a magnitude of 12.5 ohms, rather than 50 ohms. (R_{IN} would then be $20/^{40}/_{25} = 500/40 = 12.5$.)

● **Eighth-Wave Line as a Reactance**

Another use of the eighth-wavelength line is as a substitute for a physical capacitor or inductor. Specifically, when an eighth-wave line is open-terminated, its input appears as a capacitance, with reactance numerically equal to the characteristic impedance, Z_0, of the line. Conversely, when the eighth-wave line is short-circuited at its termination end, the input to the line appears as an inductance. Again, the inductive reactance is numerically equal to the characteristic impedance of the line.

A salient feature of such simulated reactance is its purity. Physical capacitors and inductors will inevitably display greater losses than their eighth-wave line counterparts. Greater precision can be implemented in locating these line reactances in the network than is readily achieved with relatively bulky capacitors and inductors. Also, microstrip construction techniques provide a considerable range of reactance values; these are governed by width of the line element and the nature of the dielectric between it and the ground plane.

When such eighth-wave lines are used in place of physical capacitors or inductors to cancel the input or output reactance of a transistor, they are known as *stub* reactances. Stub reactances are selected to present equal but opposite reactances to the parallel-equivalent reactance seen at the transistor terminal. (Note that this is different from the procedure used with eighth-wave elements as impedance-matching components; there the impedance seen at the transistor terminal was expressed in the series-equivalent format.)

Figure 5-8 shows typical deployment of eighth-wave lines as reactance stubs. In the input circuit, the eighth-wave line is open at its "far" end and therefore acts as a capacitance. Conversely, the twin output stubs terminate in RF shorts, provided by dc blocking capacitors. These stubs act as inductances. Note that by feeding in the dc collector voltage at the RF-shorted end of one of these stubs it is feasible to dispense with an RF choke in this portion of the circuit.

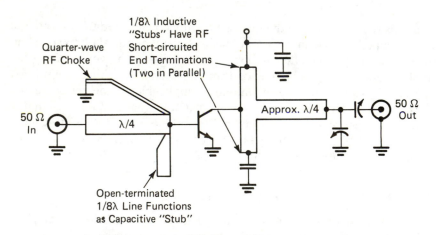

_____ FIGURE 5-8 _____
Typical Use of Eighth-Wave Stubs for Simulating Capacitance or Inductance.
The transistor is construed to require capacitive reactance to cancel its inductive input reactance and to require inductive reactance to cancel its capacitive output reactance.

● Additional Attributes and Uses of the Eighth-Wave Stubs

Although an open-ended eighth-wave stub simulates a capacitor at the operating frequency, its behavior differs markedly from the physical capacitor at the second harmonic. The reactance of the capacitor is merely one half the value pertaining to the fundamental frequency. The capacitive stub line, however, appears as a *short* to the second harmonic. That this is so can be seen by noting that such an eighth-wave line becomes a quarter-wave line at the second-harmonic frequency. And it is the nature of a quarter-wave line with its far end open to present a short circuit to the source. Inasmuch as impedance-matching networks are also exploited for their harmonic attenuating capabilities, such behavior of the eighth-wave capacitive stub is destined to have practical implications. One of these is that superior attenuation of second-harmonic energy obtains over networks using physical capacitors in their shunt arms. Therefore, the overall filtering problem is often relaxed.

It has also been found that the eighth-wave capacitive stub can be used in an output network is such a way as to significantly improve the collector efficiency of the transistor. For this to be accomplished, the first element in the output network must be a series inductance. Networks such as those shown in A, C_2, and D of Table 4-2 can be used or readily modified to be used to attain this objective. The basic idea is that the first inductive element isolates the physical collector from the first capacitive shunt element. This results in a fortuitous buildup of harmonics such that the collector voltage approaches a *square* wave shape. The transistor then functions more like an ideal switching-device, and dissipation from the simultaneous presence of both collector voltage and collector current is reduced. (At low frequencies this enhanced collector efficiency can be realized with L and tee networks using ordinary capacitors in the shunt arms. At VHF, UHF, and microwaves, the eighth-wave stub produces better results.)

Figure 5-9 shows how the eighth-wave stub is incorporated for this purpose. Its physical location along the series inductive or impedance-transforming element is best determined empirically. The ideal position of this tap appears to be on the order of $\frac{1}{16}$ wavelength from the collector of the transistor. With this technique, collector efficiencies exceeding 80% have been realized from 25-W amplifiers operating at 400 MHz.

● Other Lines for Impedance Matching

The use of quarter-wave and eighth-wave lines as impedance transformers has been described. Other transmission-line lengths can often be applied for this function. Again, however, certain criteria must be met.

Transmission-line elements somewhat shorter than one-quarter wavelength display the useful property that direct transformation can be accomplished from a purely resistive termination to one containing inductive reactance. Thus, such a line can often be used between a 50-ohm input source and the base of the driven power transistor. The mathematics of such an impedance match indicates that no stubs or other reactances are needed. In order for the match to be realized, two equations must be satisfied. They are as follows:

$$Z_0 = \sqrt{R1 \times R2} \times \sqrt{1 - \frac{(X2)^2}{R2(R1 - R2)}}$$

and

$$\tan \beta 1 = Z_0 \left(\frac{R1 - R2}{R1 \times R2} \right)$$

where Z_0 = characteristic impedance of the line
$R1$ = resistive termination of the line, usually 50 ohms
$R2$ = resistive component of the transistor impedance
$X2$ = reactive component of the transistor impedance (must be inductive)
$\beta 1$ = electrical length of the line in degrees

Some insight into these equations can be attained by supposing that we have a situation using a quarter-wave line which provides transformation between $R1$ and $R2$. In such a case, $X2$ as well as $X1$ is necessarily zero. This reduces the first equation to the simpler form, $Z_0 = \sqrt{R1 \times R2}$, which is correct for quarter-wave lines used as impedance transformers. And because $X2$ is zero, the line length corresponds to the tangent of infinity, which is 90 electrical degrees, or a quarter-wavelength. Thus, these equations are generalized relationships that embrace the quarter-wave situation.

Further consideration reveals another aspect of these equations. It can be seen that $X2$ cannot be too large, for otherwise the term $(X2)^2/[R2(R1 - R2)]$ will exceed unity and Z_0 will not be realizable. It can also be appreciated that the reactance represented by $X2$ must be inductive, not capacitive, for otherwise the calculation for line length cannot be made.

Notwithstanding these limitations, the use of such a "shortened" quarter-wave line is very desirable where applicable.

Figure 5-10 depicts a typical situation in which a line section shorter than an electrical quarter-wavelength accomplishes impedance transformation between a purely resistive 50-ohm source and the complex impedance of the base of a transistor. In a qualitative way, one can postulate that the inductive component of the transistor input impedance replaces a portion of what would otherwise be a quarter-wave line. The 0.1774 wave-

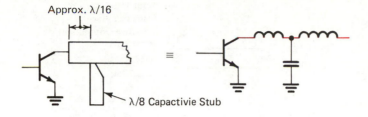

Approx. λ/16

λ/8 Capactivie Stub

≡

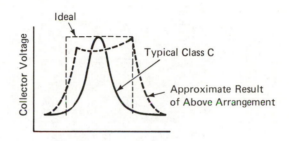

Collector Voltage

Ideal

Typical Class C

Approximate Result
of Above Arrangement

_____ **FIGURE 5-9** _____
Technique for Obtaining Increased Collector Efficiency.
By providing a small amount of inductance at collector, the
harmonic composition of the collector voltage can be tailored to
develop a closer approach to a rectangular wave.

length pertains to electrical length or to the length under ideal conditions
in air. Inasmuch as stripline construction is implied, the effective dielectric constant of the insulation material must be taken into account in the
determination of the physical length of the line. If, for example, the effective dielectric constant is 2.5, the physical length of the line will be very
nearly $0.1774/\sqrt{2.5}$ or 0.112 wavelength as measured in air. The electrical
wavelength remains 0.1774, of course.

In principle, the inductive reactance of the transistor used in this example could have been as high as approximately 17.3 ohms, but awkward
design parameters would accrue before this value was closely approached. Even less design latitude is afforded with the use of coaxial
lines because of the relatively limited characteristic impedance values
available. It is permissible to connect coaxial lines in parallel, however.
Another way of gaining design flexibility is to use twisted-pair lines,
where the characteristic impedance can be manipulated by the tightness
of the twist. Electrical losses and mechanical problems tend to occur in
twisted lines.

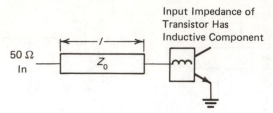

Example: Calculate line to transform input impedance
of transistor = $7 + j8\ \Omega$ to $50\ \Omega$.

$$Z_0 = \sqrt{R_1 \times R_2} \times \sqrt{1 - \frac{(X_2)^2}{R_2(R_1 - R_2)}}$$

$$= \sqrt{50 \times 7} \times \sqrt{1 - \frac{8^2}{7(50 - 7)}}$$

$$= \sqrt{350} \times \sqrt{\frac{64}{7(43)}} = 18.71 \times \sqrt{1 - 0.213}$$

$$= 18.71 \times \sqrt{0.787} = 18.71 \times 0.887$$

$= 16.6\ \Omega$, a value readily attainable with stripline techniques

Then, $\tan \beta l = Z_0 \left(\dfrac{R_1 - R_2}{R_1 \times R_2} \right) = 16.6 \left(\dfrac{50 - 7}{50 \times 7} \right)$

$$\tan \beta l = 16.6(43/350) = 16.6(0.1229)$$

$$\tan \beta l = 2.0394, \quad \beta l = \text{Arctan } 2.0394 = 63.883°$$

and

$$l = 63.883\lambda/360 = 0.1774\lambda$$

_____ **FIGURE 5-10** _____
**Direct Impedance Transformation with a "Shortened"
Quarter-Wave Line.**
Under appropriate conditions, the inductive component of the
transistor input impedance can replace a portion of a line which
otherwise would have to be one quarter of an electrical
wavelength.

● **Half-Wave Transmission Line**

A transmission line with an electrical length of one half-wavelength
behaves as an impedance *repeater*. This is shown in Figure 5-11. In Figure
5-11(a) the input impedance of the half-wave line is Z_2, the load impedance at the far-end termination. Conversely, the output impedance of the
half-wave line in Figure 5-11(b) is Z_1, the source impedance. Interest-

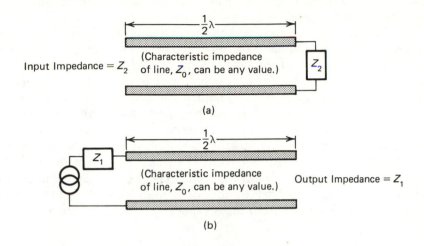

FIGURE 5-11

Impedance Repeating Feature of the Half-Wave Transmission Line.

(a) Impedance "seen" at input of half-wave line is far-end or load impedance, Z_2. (b) Impedance "seen" at the far end of half-wave line is the input impedance, Z_1.

ingly, the characteristic impedance, Z_0, of the line does *not* affect this behavior. Moreover, Z_1 and Z_2 need not be resistive, but can have any impedance value, including zero (a short-circuit) or infinity (an open-circuit). Note that the situations depicted in Figure 5-11(a) and (b) are separate and independent operating modes; one arrangement does not bear on the other.

This "transparency" of the half-wave line exists also for longer lines which are integral multiples of a half-wavelength. Thus, a full-wavelength line and $\frac{3}{2}$-wavelength lines will repeat impedances in the same fashion. If the line is many half-wavelengths, as in a long antenna feedline, the additional factor of attenuation may have to be taken into account in the delivery of power from the source to the load.

Yet another way of looking at the short-circuited half-wave transmission line is that it can simulate the action of a series-resonant circuit. In such a circuit, the source "sees" a very low impedance, ideally zero, at the resonant frequency. Conversely, the open-circuited half-wave line can simulate the action of a parallel-resonant circuit, in which case the source works into a very high impedance (ideally infinite) at the resonant frequency. Inasmuch as these resonant circuit simulations can be accomplished with quarter-wave lines, half-wave lines are not commonly found in tank circuits of amplifiers or oscillators.

A little thought will show that there is no ambiguity between half- and quarter-wave line behavior. Quarter-wave behavior manifests itself only for *odd* multiples of a quarter-wave. Thus, we can obtain similar quarter-wave behavior from lines one quarter, three quarters, five quarters, or any *odd* number of quarter-wavelengths. None of these lines can be said to be also half-wave lines. For example, a line three half-waves long is also accurately described as a six quarter-wave line, but as such is not an *odd* multiple of quarter wavelengths. Accordingly, such a line would *not* display quarter-wavelength behavior, but would behave essentially as a simple half-wavelength line. At the same time, it can be appreciated that any line can be made to display either quarter- or half-wavelength characteristics if the *frequency* from the driving source is varied sufficiently.

● Transmission-Line Connections Between Driver and Power Amplifiers

Figure 5-12 illustrates commonly encountered operating conditions when transmission lines are used between the driver and power amplifier. The flat line of part (a) is the best because a condition of impedance match prevails regardless of the length of the line or the frequency. Situation (b) also results in a flat line, but maximum available power cannot be obtained from the driver. The VSWR is unity for situations (a) and (b).

In situation (c) of Figure 5-12 there will be standing waves on the line because of reflections from the mismatched load impedance. A line length can be found wherein maximum power can be delivered to the load under these conditions. The energy transport efficiency of the system will, at best, be lower than that attainable in situations (a) and (b). Once the line length is optimized, it will not hold if the frequency is appreciably changed. Thus, for broadband work the situations depicted in (a) or (b) are very desirable.

Situation (d) of Figure 5-12 also gives rise to standing waves. Here, again, optimized energy transport is attainable at a discrete line length. In this case, however, the maximum available power can be extracted from the driver when the line length is optimum. In this respect, it is as efficient as is the operating mode of situation (a). However, the optimum operating condition is, unlike situation (a), sensitive to both line length and frequency.

Figure 5-13 shows a special case for the generalized operating mode of Figure 5-12(c). In this case, the transmission line behaves as a quarter-wave transformer, and an overall impedance match is established. Because of the transformer action, the maximum available driver power is

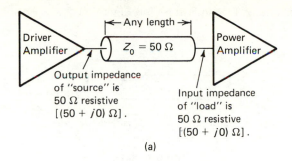

(a)

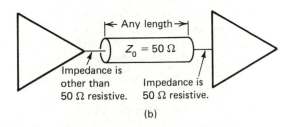

(b)

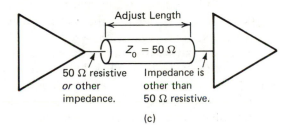

(c)

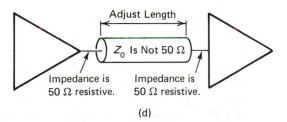

(d)

FIGURE 5-12

Commonly Encountered Line Situations in Driver-Power Amplifier Connection.

(a) Ideal situation: flat line with maximum energy transport.
(b) Line is flat, but less than maximum drive power is available.
(c) Standing waves on line: line is sensitive to both length
and frequency. (d) Standing waves on line: line is sensitive to
both length and frequency.

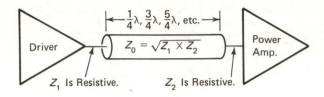

FIGURE 5-13

Special Case of Situation (c) in Figure 5-12.
Here the transmission line functions as a quarter-wave
transformer, providing an impedance match for source and load.
Energy transport is maximized, but the match is sensitive to line
length and frequency.

available for delivery to the input of the power amplifier. The efficiency is essentially as good as in Figure 5-12(a), but, of course, with sensitivity to line length and frequency. As an example, suppose that 50-ohm coaxial cable is used. Then a close impedance match can be effected if the driver output impedance is 60 ohms resistive and the power amplifier input impedance is about 41.7 ohms. That is, $Z_0 = 50 = \sqrt{60 \times 41.7}$. Many other pairs of impedance values can satisfy the basic requirement, which is $\sqrt{Z_1 \times Z_2} = Z_0$. Thus, $\sqrt{90 \times 27.78} = 50 \ \Omega$. Note that we have the option of going from a 90-ohm source to a 27.78-ohm load, or from a 27.78-ohm source to a 90-ohm load. The quarter-wave transmission-line transformer is equally adaptable whether stepping up or down.

● **Flat Line**

A *flat* transmission line is one in which no standing waves exist, that is, one operated so that the VSWR is unity. This operating condition accrues when the source impedance, the characteristic impedance of the line, and the load impedance are all the same value. Moreover, since the characteristic impedance is essentially resistive, the source and load impedances must be resistive, too. In practice, this generally involves 50-ohm coaxial cable and 50-ohm resistive source and load. Ideally, such a line constitutes the most efficient and the least trouble-prone method of feeding energy from a driver to an output amplifier or from the output amplifier to the load. One of the most useful features of the flat line is that its conveyance of energy is not frequency sensitive. An equally important corollary of this feature is that the operation of such a line is not a function of its physical or electrical length. That is, making the flat line longer or shorter will not disturb the unity VSWR. Indeed, unity VSWR is the operating condition which defines the flat line.

The "flatness" of a transmission line has nothing to do with its balance with respect to ground. Thus, both two-wire lines and coaxial-cable lines may be operated in their flat modes even though the former may be balanced and the latter, unbalanced.

● Tuned Lines

Energy may still be transported via transmission lines which are not subject to the ideal terminating conditions described previously. Generally, this comes about because of wrong impedance at the load. Counterreactance can then be inserted in such a way that the transmission line becomes part of a resonant system. Because of the high standing-wave ratio which may attend such operation, electrical losses in the line tend to be greater than in the flat line. If the line is not too long and if its dissipative losses are low, it may still operate quite efficiently in this mode. However, such a line is inherently frequency and length-sensitive. The efficiency of a system using a tuned line will be at its maximum attainable value when the resistive component of the source impedance, the characteristic impedance, and the resistive component of the load impedance are nearly equal (low VSWR). Less efficiency will result if only the resistive components of source and load are close. If even this is not the case, the efficiency of energy transport from source to load will be further impaired. The line can still be resonated, however, for optimal efficiency under these conditions.

The primary application of the tuned line is as feeder between the output amplifier and the antenna. However, with the inordinate emphasis on low VSWR, much effort is expended in making these feeder lines operate as flat as possible. They are "tuned" only in the sense that residual reactance is neutralized by the adjustment provisions. Tuned lines with high VSWR are, nonetheless, satisfactory if their higher losses, as well as frequency and length sensitivity can be tolerated.

● Transmission-Line Transformer

The transmission-line transformer is a unique and useful RF component. Its physical resemblance to a conventional transformer is deceiving, for its operating principle is quite different. And its performance, particularly its efficiency and broadband response, clearly indicate that more than meets the eye is involved. A common physical form consists of a few turns of paired wires on a toroidal core. Casual inspection would suggest it to be an ordinary bifilar wound transformer. The winding connections might be suggestive of an autotransformer.

Closer involvement with such a transformer could turn up some strange features. It would be discovered that the magnetic core is nearly passive as a means of energy transfer between input and output terminals. And, mysteriously, one would see negligible degradation in high-frequency response from interwinding capacitance. Further investigation might reveal that the number of turns do not stem from classical transformer design equations. And, if one has had considerable experience with ordinary transformers, surprise at the high power levels a compact device of this nature can handle, all the while running cool, would be natural. Finally, it will also be seen that there are very low level nonlinear effects even when operating at high power.

The preceding facts, and more, are attributable to the fact that the "windings" do not function as inductively coupled coils, but rather as short transmission lines; generally at the highest frequency of operation, these lines are still less than an eighth-wavelength. The coiled or helical format of these lines has no direct electrical significance because a transmission line can be coiled up without significantly altering its electrical characteristics. (This is especially true for coaxial cable, but for practical purposes can be said to be substantially true of other line formats, such as twisted wire, or closely paired wire.) Coiling simply compacts the physical dimensions of the lines. In so doing, however, it indirectly contributes to the electrical realization of the device, for it permits certain terminals to be connected together almost directly.

Figures 5-14 and 5-15 show some common versions of transmission-line transformers. In all these the characteristic impedance of the line, regardless of its type, is the geometric mean of the input and output resistances; that is, $Z_0 = \sqrt{R_{IN} \times R_{OUT}}$. Inasmuch as these are all four-to-one or one-to-four impedance transforming devices, we have by substitution

$$Z_0 = \sqrt{R \times \frac{R}{4}} = \sqrt{\frac{R^2}{4}} = \frac{R}{2}$$

Thus, Z_0 is selected or fabricated to be one-half the value of the larger terminating resistance. Or, stated another way, Z_0 is twice the value of the smaller terminating resistance. Here we see a difference in the application of these devices from that of conventional transformers, where considerable latitude pertains to the terminating impedances. In actual practice, the three impedances involved (R_{IN}, R_{OUT}, and Z_0) in the application of the transmission-line transformer often deviate somewhat from the mathematical ideal. Surprisingly, the bandwidth and the transformation ratio will often remain within acceptable limits.

Both cored and coreless devices are shown in the several illustrations. It must not be supposed that the cores "load" the transmission lines, for neither the characteristic impedance nor the electrical wavelength are affected by the magnetic cores. Moreover, those cores do not participate in the transfer of energy as in a conventional transformer. It will be observed that the core material is not inserted between the line

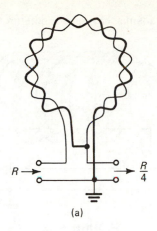

$R \rightarrow$ $\dfrac{R}{4}$

(a)

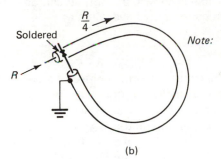

Soldered

$\dfrac{R}{4} \nearrow$

$R \rightarrow$

Note: Input and output are with respect to indicated ground. Do not ground cable elsewhere.

(b)

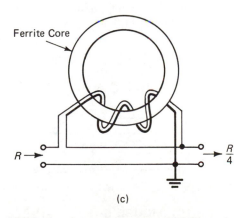

Ferrite Core

$R \rightarrow$ $\rightarrow \dfrac{R}{4}$

(c)

FIGURE 5-14

**Three Simple Implementations of Transmission-Line
Transformers.**

(a) Twisted Wire. (b) Coaxial cable. (c) Paired wire on ferrite core.
(Twisted wire or coaxial cable are also used in this fashion.)

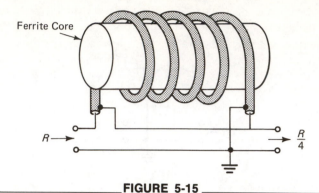

Ferrite Core

$R \longrightarrow$ $\dfrac{R}{4}$

FIGURE 5-15
Transmission-Line Transformer Wound on Cylindrical Core.
Although coaxial cable is shown, this format can be implemented
with paired or twisted wires as well.

elements, but rather adjacent to them. The use of a core does not significantly alter performance at higher frequencies, but appreciably extends
the low-frequency operating ability. Let us see why this is so.

Figure 5-16 is the usual schematic diagram used to represent the
transmission-line transformers of Figures 5-14 and 5-15. The coil symbols
should not be construed to imply inductive coupling between the transmission-line elements, as between primary and secondary of conventional
transformers. However, this symbology is useful in depicting that the line
elements, in addition to behaving as a transmission line, also exhibit ordinary inductance from end to end. This is important because one line ele-

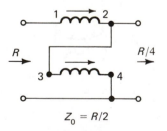

R

$R/4$

$Z_0 = R/2$

FIGURE 5-16
**Schematic Diagram of the Transmission-Line Transformers of
Figures 5-14 and 5-15.**
Such a representation is useful in depicting the terminal
connections, but should not be construed to imply the kind of
inductive coupling associated with ordinary transformers.

ment will be seen to be shunted directly across either the source or the load. We must ask ourselves how this situation can prevail without having a short-circuit.

The answer is that a line element, such as the one symbolized by "inductor" 3-4 of Figure 5-16 does indeed present ordinary inductive reactance to the flow of such shunt current. The essential thing to grasp here is that this inductive reactance stems from this line element's existence as a conductor. It has nothing to do with this line element's participation with line element 1-2 to form a transmission line. This being the case, it follows that the low-frequency shunting action of line element 3-4 would be reduced if its *inductance* could be increased. Then we should expect greater low-frequency response from our transmission-line transformer. But how can this be brought about without changing the characteristics of the transmission line itself? One way is to coil up the transmission line so that its shape is, indeed, that of a coil. This expedient, however, is generally necessary anyway in order that the terminal connecting wires be short. If, furthermore, the line is coiled around ferrite or other suitable magnetic material, the inductance of the single elements will be greatly increased, but there will be virtually no effect on the transmission line parameters. (As pointed out previously, the ferrite would have to be inserted *between* the line elements in order to affect the characteristic impedance or the electrical line length.)

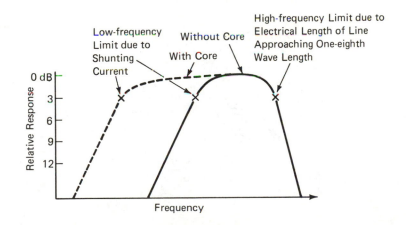

FIGURE 5-17

General Effect of Adding a Magnetic Core to a Transmission-Line Transformer.

Low-frequency response is extended because more inductive reactance is provided to impede the flow of shunting currents.

From what has been described, the desired effect of the ferrite core is to make such a single line element as 3-4 a more effective "RF choke" to low-frequency shunt currents. Although early literature suggested transmission-line action at high frequencies, and ordinary transformer action at low frequencies, present thinking tends to ascribe operation in the transmission-line mode throughout the passband of the device. The core merely *extends* the low-frequency response via the described mechanism. Figure 5-17 shows the general effect of a core. In practice it is observed that there is much less tendency for temperature rise in the core compared to the situation with an ordinary transformer operating under similar power conditions. This reinforces the theory that the core does not participate significantly in the transfer of energy from input to output. A corollary of this is that the core need not be made of exotic high-frequency material, and can be of smaller volume than in a conventional transformer with similar ratings.

● One-to-One Transmission-Line Transformer

The arrangement shown in Figure 5-18 does not transform impedance values but exhibits the interesting property of functioning as a transposition device between balanced and unbalanced circuitry. That is, it is inherently a *balun*. At first inspection, it would appear that this arrangement

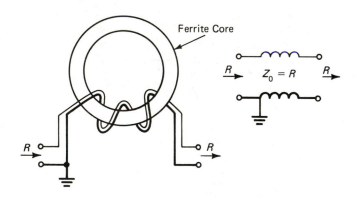

_____ **FIGURE 5-18** _____
One-to-One Transmission-Line Transformer and Its Schematic Diagram.
The primary use of this arrangement is to convert from unbalanced to balanced circuitry, or vice versa.

differs from the four-to-one configuration of Figure 5-16 only in the connection of its leads. Qualitatively, this is so. However, the characteristic impedance in this case is R, rather than $R/2$ ohms. (In both of these transmission-line implementations, Z_0 is the geometric mean of the input and output impedances. Thus, if the input or output impedance changes because of a different transformation ratio, it follows that a different Z_0 must be used.)

A definite resemblance exists between this transmission-line balun and the bifilar-wound RF choke used in the filament circuit of grounded-grid tube amplifiers. The operation of the latter component is predicated on its high inductance, and any "characteristic impedance" mutual to the windings plays no direct part in providing high impedance to RF. Moreover, such a choke is used as a shunt element with respect to the RF. Conversely, the one-to-one transmission-line transformer must have a discrete characteristic impedance, and is used as a series element to transport RF with minimal rather than with high attenuation. Although its inductance is important in broadbanding its performance, inductance is not the basic parameter underlying its operation, as it is in the bifilar-wound RF choke.

● Transmission-Line Transformers for Other Impedance Ratios

If multiple line elements are properly interconnected, other impedance transformation ratios than 4 to 1 or 1 to 4 can be realized. Figure 5-19 illustrates the technique for devising transmission-line transformers with impedance ratios of 9 to 1 (or 1 to 9) and 16 to 1 (or 1 to 16). Although the general scheme can be extended to ratios of 25 and 36, practical difficulties of construction tend to make the higher ratios somewhat difficult to implement.

In its simplest format, these transmission-line transformers consist of twisted wires: three for the 9-to-1 transformer and four for the 16-to-1 transformer. Despite the multiplicity of turns shown in the schematic diagrams, these transformers often utilize a single turn, or slightly less than a single turn of the twisted wires. This stems from the rule that the length of the wires should be no more than about one eighth-wavelength at the highest frequency of interest. If desired, a ferrite core can be used in order to extend the transformer's low-frequency operation. Generally, some empirical work must be done to obtain optimum results, particularly if broadbanding is the objective. For example, some experimentation is usually needed to determine the way in which the wires are twisted together, for this directly governs the characteristic impedance of adjacent wires. Fortunately, the parameters are not critical for most applications,

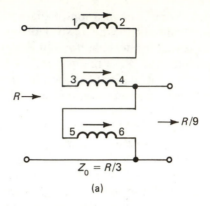

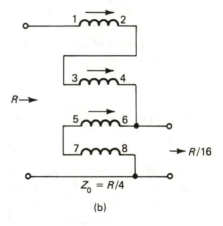

FIGURE 5-19
**Diagrams of 9 to 1 (or 1 to 9) and 16 to 1 (or 1 to 16)
Transmission-Line Transformers.**
(a) Arrangement for impedance transformation by factor of 9 or
1/9. (b) Arrangement for impedance transformation by factor of 16
or 1/16.

and it is usually easy to attain better results than with a conventional elec-tromagnetic-type transformer.

As would be expected, the phasing of the line elements is all-impor-tant. The enumeration shown in the diagrams of Figure 5-19 corresponds to the ends of the twisted wires. Even numbers correspond to one end; odd numbers correspond to the opposite end.

Various constructions can actually be used. Instead of twisted wires, parallel wires or multifilar winding formats are applicable. And either a solenoidal or toroidal core may merit consideration. As with the 4-to-1 line-transformer previously described, the basic operating principle is premised upon the conductors behaving as short transmission lines. Accordingly, it can prove misleading to think of the "windings" as being inductively coupled in the manner of conventional electromagnetic transformers. That this is not so is readily inferred by the relatively cool operation of a core, if used, even when handling high power levels. Eddy current and hysteresis loss in the core are minimal because, unlike in the conventional transformer, the core does not participate in the energy transfer between the "windings."

● Special Considerations When Designing with Stripline Elements

The simple formulas found in handbooks for calculating the characteristic impedance of two-wire and coaxial transmission lines are not generally applicable to the stripline structure. The primary reason is that this structure gives rise to complex fringing patterns of the electric lines of force and to proximity effects. This, in turn, results in an effective value of the relative dielectric constant which is *lower* than would be measured by incorporating the board material in a capacitor "sandwich" in which both copper plates have extensive and equal areas. In the stripline structure, as shown in Figure 5-20, the ratio *W/H* exerts a strong effect on the effective value of the relative dielectric constant. In all transmission lines, the characteristic impedance is governed by *both* geometric factors and by the relative dielectric constant. Inasmuch as the geometry of the line

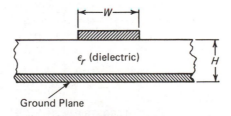

―――――――――――――――― **FIGURE 5-20** ――――――――――――――――
Basic Physical and Electrical Parameters in Stripline (or Microstrip) Circuit Elements.
Before determining the length of the line, the above parameters must be taken into account because they influence the characteristic impedance and the velocity factor.

directly influences the characteristic impedance via physical dimensions, and *also* by its effect on the effective value of the relative dielectric constant, it is evident that the mathematical format connecting Z_0 with geometrical dimensions will incline more toward complexity than simplicity.

It is also true in all transmission lines that the velocity factor is a function of the relative dielectric constant. For most practical purposes, the velocity factor in air is very close to 1.00. The insertion of

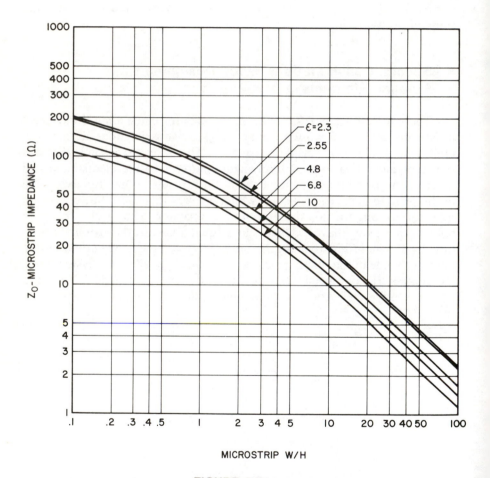

MICROSTRIP W/H

FIGURE 5-21

Characteristic Impedance of Microstrip Transmission Line versus W/H.

By appropriate selection of the dielectric material and by design of the geometric features of the line, a wide range of Z_0 values may be realized.

(Courtesy of Communications Transistor Corp.)

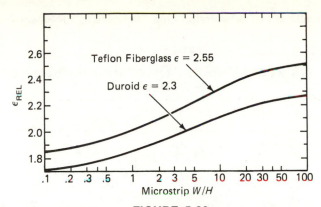

FIGURE 5-22

Effective Value of Dielectric Constant Versus *W/H* for Low Dielectric Constant Materials.

For high values of *W/H* the effective dielectric constant approaches the "rated" dielectric constant of the material.
(Courtesy of Communications Transistor Corp.)

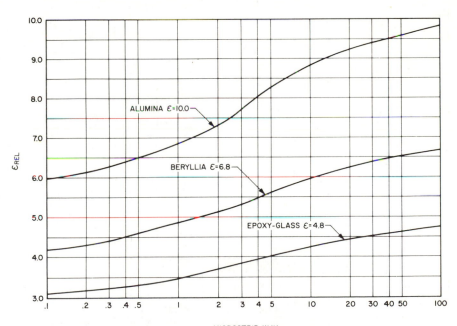

MICROSTRIP W/H

FIGURE 5-23

Effective Value of Dielectric Constant Versus *W/H* for High Dielectric Constant Materials.

A nonsimple relationship is suggested by irregularity of the curves. Nonetheless, the "rated" dielectric constant of the material is approached at high values of *W/H*.
(Courtesy of Communications Transistor Corp.)

-- TABLE 5-1 --

Microstrip Z_0 and Velocity Factor Versus Width-to-Height (W/H) Ratio[a]

W/H	Air $\varepsilon_R = 1.0$		Teflon $\varepsilon_R = 2.55$		Epoxy $\varepsilon_R = 4.25$		Alumina $\varepsilon_R = 9.6$	
	Z_0	V_P	Z_0	V_P	Z_0	V_P	Z_0	V_P
0.630	168.425	1.000	110.683	0.657	87.986	0.522	60.977	0.362
0.695	161.878	1.000	106.258	0.656	84.414	0.521	58.441	0.361
0.766	155.370	1.000	101.865	0.656	80.870	0.521	55.927	0.360
0.844	148.909	1.000	97.509	0.655	77.360	0.520	53.440	0.359
0.931	142.506	1.000	93.199	0.654	73.888	0.518	50.985	0.358
1.026	136.171	1.000	88.941	0.653	70.463	0.517	48.566	0.357
1.131	129.916	1.000	84.745	0.652	67.090	0.516	46.187	0.356
1.247	123.753	1.000	80.616	0.651	63.775	0.515	43.853	0.354
1.375	117.692	1.000	76.565	0.651	60.524	0.514	41.568	0.353
1.516	111.746	1.000	72.597	0.650	57.345	0.513	39.337	0.352
1.672	105.926	1.000	68.721	0.649	54.243	0.512	37.164	0.351
1.843	100.242	1.000	64.944	0.648	51.223	0.511	35.053	0.350
2.032	94.706	1.000	61.273	0.647	48.291	0.510	33.007	0.349
2.240	89.327	1.000	57.714	0.646	45.451	0.509	31.030	0.347
2.470	84.115	1.000	54.271	0.645	42.709	0.508	29.123	0.346
2.723	79.076	1.000	50.951	0.644	40.066	0.507	27.289	0.345
3.002	74.218	1.000	47.757	0.643	37.527	0.506	25.531	0.344
3.310	69.546	1.000	44.692	0.643	35.094	0.505	23.849	0.343
3.649	65.065	1.000	41.759	0.642	32.768	0.504	22.244	0.342
4.023	60.779	1.000	38.959	0.641	30.550	0.503	20.716	0.341
4.435	56.689	1.000	36.292	0.640	28.440	0.502	19.266	0.340
4.890	52.796	1.000	33.760	0.639	26.439	0.501	17.892	0.339
5.391	49.100	1.000	31.360	0.639	24.544	0.500	16.594	0.338
5.944	45.600	1.000	29.091	0.638	22.755	0.499	15.370	0.337
6.553	42.291	1.000	26.952	0.637	21.069	0.498	14.218	0.336
7.224	39.173	1.000	24.938	0.637	19.485	0.497	13.138	0.335
7.965	36.233	1.000	23.047	0.636	17.998	0.497	12.125	0.335
8.781	33.484	1.000	21.275	0.635	16.606	0.496	11.179	0.334
9.681	30.904	1.000	19.618	0.635	15.305	0.495	10.295	0.333
10.674	28.491	1.000	18.071	0.634	14.091	0.495	9.472	0.332
11.768	26.240	1.000	16.629	0.634	12.961	0.494	8.707	0.332
12.974	24.143	1.000	15.288	0.633	11.911	0.493	7.996	0.331
14.304	22.192	1.000	14.043	0.633	10.937	0.493	7.338	0.331
15.770	20.381	1.000	12.888	0.632	10.033	0.492	6.728	0.330
17.387	18.702	1.000	11.818	0.632	9.198	0.492	6.164	0.330
19.169	17.148	1.000	10.830	0.632	8.425	0.491	5.644	0.329
21.133	15.172	1.000	9.917	0.631	7.713	0.491	5.164	0.329
23.300	14.385	1.000	9.074	0.631	7.056	0.490	4.722	0.328
25.688	13.162	1.000	8.299	0.630	6.451	0.490	4.315	0.328

[a] *Courtesy of TRW*

TABLE 5-1 (Cont'd.)

Table continued

W/H	Air $\epsilon_R = 1.0$		Teflon $\epsilon_R = 2.55$		Epoxy $\epsilon_R = 4.25$		Alumina $\epsilon_R = 9.6$	
	Z_o	V_p	Z_o	V_p	Z_o	V_p	Z_o	V_p
28.321	12.036	1.000	7.585	0.630	5.894	0.490	3.942	0.327
31.224	10.999	1.000	6.929	0.630	5.383	0.489	3.598	0.327
34.424	10.047	1.000	6.326	0.630	4.914	0.489	3.284	0.327
37.953	9.172	1.000	5.773	0.629	4.483	0.489	2.995	0.327
41.843	8.370	1.000	5.266	0.629	4.089	0.489	2.731	0.326
46.132	7.634	1.000	4.801	0.629	3.727	0.488	2.489	0.326
50.860	6.960	1.000	4.376	0.629	3.397	0.488	2.267	0.326
56.073	6.343	1.000	3.987	0.629	3.094	0.488	2.065	0.326
61.821	5.779	1.000	3.632	0.628	2.818	0.488	1.880	0.325
68.157	5.264	1.000	3.307	0.628	2.566	0.487	1.711	0.325
75.144	4.792	1.000	3.010	0.628	2.335	0.487	1.557	0.325
82.846	4.362	1.000	2.739	0.628	2.125	0.487	1.417	0.325
91.337	3.969	1.000	2.492	0.628	1.933	0.487	1.289	0.325
100.700	3.611	1.000	2.267	0.628	1.758	0.487	1.172	0.324

dielectric material other than air decreases both the characteristic impedance and the velocity factor. The practical effect of reduced velocity factor is that, for a given electrical length, the physical length of the line will *not* be the same as it would be with air dielectric. Specifically, the physical length will be reduced by the factor $\dfrac{1}{\sqrt{\epsilon_r}}$ where ϵ_r is the effective value of the relative dielectric constant. This, in most cases, is a favorable situation, for it enables more compact construction. For example, a quarter-wavelength line which would require, say, 10 centimeters of air dielectric transmission line (dielectric constant = 1.00) would require a length of only 5 centimeters if constructed with a dielectric material with a relative dielectric constant of 4. Also, the new velocity factor would be 0.5 rather than unity.

Figure 5-21 depicts characteristic impedance, Z_0, as a function of the ratio W/H for striplines utilizing several different dielectrics. Figures 5-21, 5-22 and 5-23 show the variation in the effective value of the relative dielectric constant. It will be noted that for high W/H ratios the effective value of the relative dielectric constant approaches that of the material itself. Similar information is given in tabular form in Table 5-1 in which the velocity factors are also indicated. The use of such curves and tables greatly simplifies the design of stripline elements.

6

Low-Power
Applications

In this chapter, examples of solid-state RF applications involving power levels of a fraction of a watt to 25 watts (W) are discussed. Within this range, one finds greater similarity to tube-circuit practice than is the case at higher power levels. Also, it is often the case at VHF, UHF, and microwaves that a few watts suffice for adequate performance. Yet another important aspect of low-power circuits is that the oscillators, buffers, drivers, and frequency multipliers which precede higher-power final amplifiers operate at much lower power levels. Thus, low-power RF stages are of interest both as RF output circuits and as building blocks of more elaborate systems.

   The philosophy underlying the circuit descriptions is that the reader will most likely establish relevancy with a somewhat similar, but not identical, application. Therefore, the editorial involvement is with basic principles rather than with detailed "how-to-build" instructions.

● Microwave Doppler-Radar System

The CW transmit-receive system shown in Figure 6-1 can be applied for speed measurement or intruder-detection functions. Because of the doppler effect, a moving object alters the frequency of the reflected energy. The received and transmitted frequencies then heterodyne in the nonlinear characteristic of the oscillator diode, thereby producing an intermediate frequency proportional to the speed of the object. (A true calibration of the object's speed is usually based upon the premise that the object is moving directly along the line of sight, either toward or away from the radar. Police radar detection of speeding autos is a classic example of this principle.)

The bulk-effect diode makes use of a negative-resistance region which develops in gallium arsenide semiconductor material when it is suitably biased by a direct current. The oscillation frequency is primarily governed by the associated cavity. However, it is also important that the power supply be ripple-free and well regulated in the interest of optimum frequency stability.

The diode used in this doppler radar is integrally associated with its own oscillator cavity and operates at 10,525 MHz (see Figure 6-2). It is only necessary to attach a horn and connect the bias source and a suitable audio amplifier. If the audio passband extends from 100 Hz to 3 kHz, the output tone will represent speeds ranging from approximately 3.33 to 100 miles per hour. Either analog or digital circuit techniques can be used to produce an appropriate meter read-out or to actuate an alarm.

Note that in this application the horn antenna is effective for both outgoing and incoming microwave energy. Practical horns can provide 16.5-dB gain. One can, therefore, deal with an effective radiated power of

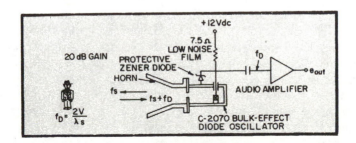

_____ **FIGURE 6-1** _____
**Speed Measurement or Intrusion-Detection Radar
Using Bulk-Effect Diode.**
The diode and cavity are available as an integral unit.
(Courtesy of General Electric Microwave Devices Products Section)

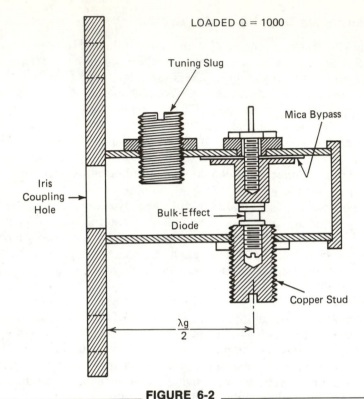

LOADED Q = 1000

Tuning Slug

Mica Bypass

Iris
Coupling
Hole

Bulk-Effect
Diode

Copper Stud

$$\frac{\lambda g}{2}$$

FIGURE 6-2
Detailed View of Diode-Cavity Assembly.
Operation is readily attained by attaching a horn and connecting
bias source and audio amplifier.
(Courtesy of General Electric Microwave Devices Products Section)

some 33 dB greater than would accrue from coupling the oscillator to a simple dipole radiator in an ordinary transmitting application. This is a typical example of how low power levels can be put to practical use in microwave applications. Working distances up to several hundred feet are feasible. The reflecting characteristics of the target and the noise characteristics of the oscillator diode are the main distance-limiting factors. The gain of the audio amplifier is important, too, but high gain is only useful if the doppler IF signal is not masked by oscillator noise.

● One-Megahertz JFET Crystal Oscillator

The JFET circuit shown in Figure 6-3 is a general-purpose crystal oscillator. Its salient feature is simplicity. It is useful where much better stability than is readily forthcoming from self-excited oscillators is desired.

On the other hand, it is not a recommended circuit for applications where the best potentialities of crystal control are sought. Although a P-channel JFET is shown, N-channel devices will work equally well providing the polarity of the supply voltage is reversed. This circuit is the JFET version of the Miller oscillator of former popularity with tubes. By making appropriate changes in the *LC* resonant tank, other frequencies can be generated with appropriate fundamental-frequency crystals. Unreliable starting performance can usually be remedied by connecting an external feedback capacitor of a few picofarads from drain to gate.

Inasmuch as crystal oscillators are generally followed by one or more amplifier stages, the low power output of small JFET's need not be a disadvantage. In any event, a buffer amplifier is very desirable in order to reduce loading and to minimize frequency pulling. Unfortunately, there is no internal buffering such as existed in tetrode and pentode tube versions of the circuit. Nonetheless, it is a commonly encountered workhorse and is a satisfactory performer in a variety of applications. In particular, the high-impedance gate circuit enables it to support the parallel-resonant mode of oscillation in medium-frequency crystals. For best results, the subsequent buffer stage should also be a JFET. Although the output is available directly at the drain terminal, as shown, it will generally be better to use an approximate mid-tap on the inductor in conjunction with a small series capacitor.

1 MHz Oscillator

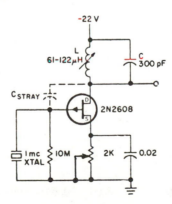

FIGURE 6-3

JFET Version of the Miller Crystal-Oscillator Circuit.

This general circuit has served as a workhorse for numerous applications where the demands for stability and frequency precision are moderate.

(Courtesy of Motorola Semiconductor Products, Inc.)

● JFET Frequency Doubler

The JFET frequency-doubling circuit of Figure 6-4 is designed for rather low power operation, but there is no reason why the power level cannot be scaled up somewhat via appropriate device selection and higher supply voltage. The operating efficiency and wave purity of this doubler is much higher than is readily attainable from conventional frequency multipliers. It will be noted that the input circuit is push-pull, whereas the output circuit is parallel. Such a configuration favors the support of even-order harmonics in the output and, at the same time, tends to cancel odd harmonics. Inasmuch as the fundamental frequency is, itself, an odd harmonic, relatively little wave-purity contamination can occur from feedthrough of the fundamental in this doubler.

The most interesting aspect of this doubling scheme has to do with the mechanism of second-harmonic production. This is due to the unique transfer characteristic of JFET's. Specifically, it approaches a true square-law response. The use of the grounded-gate connection for the JFET's resulted in more practical resonant-circuit components, but the experimenter can adapt the basic idea for grounded-source operation. As shown, input frequencies in the vicinity of 60 MHz are doubled. The output trap, L_2, C_8 attenuates residual third-harmonic energy in the output. By means of this technique, together with matched JFET's, third-harmonic rejection can be as high as 70 dB below $2f$. However, many practical applications will allow the trap to be dispensed with, for 50-dB rejection of $3f$ can still be obtained.

Unlike most active-device frequency multipliers, the doubled frequency in this system represents the main distortion product in the JFET's. That is, $2f$ in the output is not produced by shock-excitation of the resonant output circuit. To optimize such second-harmonic distortion, variable resistance R_1 is included so that the most effective bias can be empirically determined. Because of the range provided by R_1, the pinch-off voltage of the selected JFET's is not of great consequence. In any event, R_1 is adjusted to optimize the frequency-doubling characteristic of the circuit. The balancing potentiometer, R_2, is adjusted for minimum presence of the fundamental frequency in the output. To a considerable extent, R_2 will compensate inequalities in the two JFET's.

● Class C Amplifier: 1.5 Watt, 50 Megahertz

An interesting aspect of the amplifier shown in Figure 6-5 is its configurational resemblance to class C tube amplifiers. For example, the variable resistance, R, serves the same function as the "grid leak" or grid return resistance in a self-biased tube circuit. And C_6 is the counterpart of the "grid capacitor" in such circuits. That is, under drive conditions, a

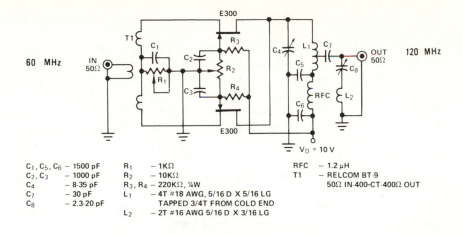

60 MHz   IN 50Ω   T1   120 MHz   OUT 50Ω

E300
$V_D = 10$ V

C_1, C_5, C_6 — 1500 pF	R_1 — 1KΩ	RFC — 1.2 μH
C_2, C_3 — 1000 pF	R_2 — 10KΩ	T1 — RELCOM BT-9
C_4 — 8-35 pF	R_3, R_4 — 220KΩ, ¼W	50Ω IN-400-CT-400Ω OUT
C_7 — 30 pF	L_1 — 4T #18 AWG, 5/16 D X 5/16 LG	
C_8 — 2.3-20 pF	TAPPED 3/4T FROM COLD END	
	L_2 — 2T #16 AWG 5/16 D X 3/16 LG	

_____ **FIGURE 6-4** _____

High-Efficiency JFET Frequency Doubler.

The JFET transfer characteristic and the circuit configuration
contribute to enhancement of second-harmonic production.
(Courtesy of Siliconix Corp.)

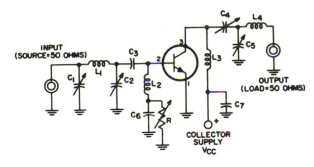

INPUT (SOURCE=50 OHMS)

OUTPUT (LOAD=50 OHMS)

COLLECTOR SUPPLY Vcc

C_1: 70–350 pf
C_2, C_4, C_5: 7–100 pf
C_3: 0.01 μf
C_6: 0.002 μf
C_7: 0.02 μf

R: 1000 ohms, variable

L_1: 0.13 μh, 4 turns, No. 18 wire,
¼" ID, closely wound

L_2: 2.4 μh, choke,
Miller Part No. 4606

L_3: 0.6 μh, 10 turns, No. 18 wire,
⅜" ID, closely wound

L_4: 0.6 μh, 10 turns, No. 18 wire,
⅜" ID, closely wound

_____ **FIGURE 6-5** _____

Class C Amplifier: 1½ Watt, 50 MHz.

This simple circuit provides a convenient way to explore solid-state
VHF techniques.
(Courtesy of RCA Solid-State Division)

191

reverse-bias charge is trapped in C_6, and this charge biases the transistor well into the class C operating region. In practice, the drive power and R can be mutually adjusted to achieve maximum RF output from the amplifier. Although this will correspond to a value of R greater than zero, such biasing is not generally encountered in transistor RF amplifiers, at least not in those with greater power capability than this amplifier. The reason is that most RF power transistors become more vulnerable to catastrophic breakdown as reverse bias is increased. However, the technique is feasible for low-power output amplifiers and drivers. (Probably 5 W or so is the level at which it may not be wise to operate deeper into class C than naturally ensues from directly grounding the base RF choke.) Aside from endangering the transistor, this biasing technique often causes the emitter-base diode of the transistor to go into Zener or avalanche breakdown, thereby defeating the self-bias function and sometimes causing loading problems in the driver stage. In this amplifier, the base RF choke, L_2, has a low Q. This discourages low-frequency "parasitic" oscillation.

The RCA 2N3118 transistor can be safely used in this circuit with a 40-V dc supply. Under such conditions, 60 mW of drive power will suffice for 1½-W output power. With 120 mW of drive, 2-W output can be approached. Sufficient heat sinking should be applied to the TO-5 case to keep it near 25°C during operation.

The internal feedback capacitance of this transistor is sufficiently low so that neutralization is not needed. However, this does not obviate the usual precautions pertaining to short leads, input-output isolation, and effective bypassing. With respect to the latter, it may prove profitable to parallel C_7 with both a smaller and a larger ceramic capacitor.

This amplifier can be readily adapted for either CW or FM service in the amateur 6-meter band. It is not, however, suitable for SSB, and lacks peak voltage and power capability to be amplitude modulated.

● JFET Broadband Linear Amplifier

The inordinately simple amplifier circuit shown in Figure 6-6 is "linear" in the sense that it is eminently suitable for processing amplitude-modulated signals. It is intended for insertion of a 75-ohm system, such as is commonly used in CATV and MATV systems. However, it is likely that little or no modification would be needed for satisfactory 50-ohm operation. Although the input, output, and interstage networks border on the primitive, the response is within ± 1 dB throughout the 200- to 250-MHz band. The low input VSWR accrues from the fact that JFET's can display virtually no input reactance when connected in the grounded-gate configuration.

Some experimentation is needed to find the optimum taps on inductors L_2, L_4, and L_6. Center-tapping is a good start for such empirical in-

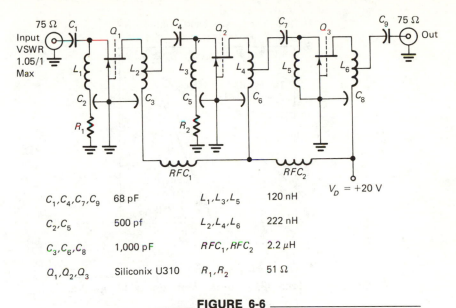

C_1, C_4, C_7, C_9	68 pF	L_1, L_3, L_5	120 nH
C_2, C_5	500 pf	L_2, L_4, L_6	222 nH
C_3, C_6, C_8	1,000 pF	RFC_1, RFC_2	2.2 μH
Q_1, Q_2, Q_3	Siliconix U310	R_1, R_2	51 Ω

FIGURE 6-6

JFET Linear Amplifier for the 200- 250-MHz band.
Simplicity of construction is the salient feature of this
broadband amplifier.
(Courtesy of Siliconix)

vestigation. Reasonably good performance will be immediately forthcoming, and the taps are not critical. Conspicuous by their absence in this circuit are resonating capacitors. Also missing, but not missed, are bias networks or bias sources. (Some effort should be directed toward placing a shield partition between source and drain leads. This prevents positive feedback in the manner of ultraaudion or Colpitts oscillators.)

The intermodulation and power-level performance of this JFET amplifier are graphically depicted in Figure 6-7.

● **Power MOSFET Driver for a Broadband Linear Amplifier System**

The circuit shown in Figure 6-8 is intended as a basic "building block" for a more elaborate linear amplifier. In particular, it would be suitable for driving a power MOSFET output stage. The adage "a chain is no stronger than its weakest link" appropriately applies to a lineup of linear-amplifier stages. If a driver stage is nonlinear and thereby develops high intermodulation distortion, all design efforts expended on the output stage for the sake of system linearity are in vain. This driver attains its low

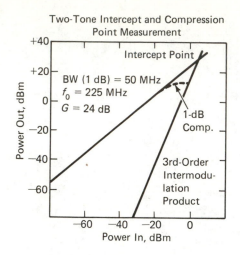

Two-Tone Intercept and Compression
Point Measurement

FIGURE 6-7

Performance of the JFET Linear Amplifier of Figure 6-6.
(Courtesy of Siliconix)

intermodulation distortion by virtue of the inherent transfer linearity of power MOSFETs when properly biased. As can be seen, provision is made for the introduction of the bias voltage, V_{GS}. Although a variable source of positive voltage is needed (approximately 4 V), virtually zero bias current is consumed by the gate.

The transmission-type transformer, $T1$, resembles types commonly used as baluns for converting from balanced to unbalanced circuits, or vice versa. However, in this instance, it is the impedance transforming property of $T1$ that is exploited. Specifically, the 50-ohm input impedance is transformed to one-fourth of this value, 12.5 ohms, in order to achieve a reasonable impedance match at the gate throughout the 40- to 265-MHz range. (The near-infinite input impedance often cited as a feature of the power MOSFET prevails only at dc and for very low frequencies.) Another factor in producing the low gate impedance is the LR feedback network between the output and the gate.

The inordinately broad response of this amplifier accrues from the combined action of transformer T1 and the LR feedback network. The 0.15-μH feedback inductor can be made by winding about seven turns of number 30 AWG enamel wire on a ½-W, 1-MΩ resistor. Molded inductors will probably have too much distributed capacity to be useful for this function. The effect of such an inductor will be to reduce the upper-frequency bandwidth. The gain and intermodulation performance of this circuit is depicted in Figure 6-24, along with that of a two-transistor version.

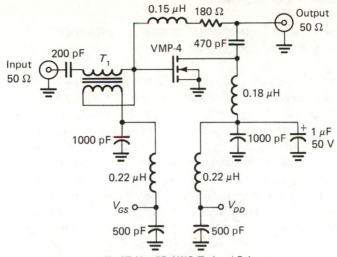

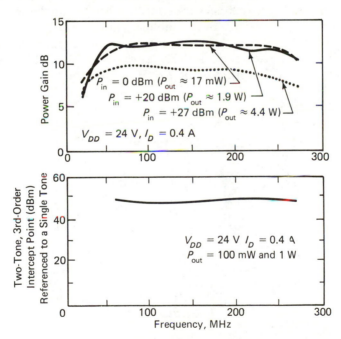

T_1 4T No. 22 AWG Twisted Pair on
Indiana General F625-902

FIGURE 6-8

**Simple 2-Watt Broadband Driver for a Linear Amplifier
System.**

Although no tuned circuits are used, the intermodulation and
harmonic distortion are acceptably low.
(Courtesy of Siliconix Corp.)

195

● Low-Power Two-Meter Linear Amplifiers Using Power MOSFETs

Simplicity, together with respectable performance, characterizes the two-meter linear amplifier shown in Figure 6-9. Simple lumped-circuit resonant circuits are used, and there is no need for neutralization or feedback networks. R_5 enables adjustment of the bias for optimum linearity. Unlike a bipolar transistor, the power MOSFET does not consume dc bias current. To promote stable and reproducible operation, the bias supply uses a 12-V Zener diode. Input and output impedances are 50 ohms. The power MOSFET can withstand any VSVR, so no precautions are necessary during either adjustment or operation. Considerable inherent protection from overdrive or overloading exists because the output current decreases with rising temperature. Inasmuch as the case of the device is internally connected to the drain, the heat sink must be insulated from the ground plane of the circuit board. This produces the capacitor, C_{10}. The VN66AJ power MOSFET can be used with substantially the same results.

Another version of this basic amplifier is shown in Figure 6-10. Here two power MOSFETs contribute nearly twice the output power available from the single device amplifier. The two power MOSFETs are essentially in parallel with respect to the RF. However, they are not connected

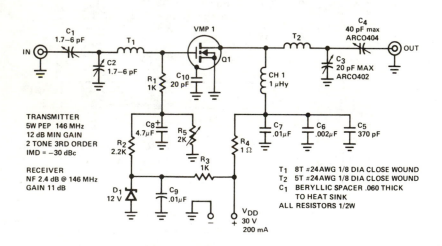

_____ **FIGURE 6-9** _____

Two-Meter, 5-Watt PEP Linear Amplifier Using a Power MOSFET.

Although intended as a transmitter output amplifier, the very same circuit surprisingly performs well as an RF preamplifier for a receiver.

(Courtesy of Siliconix Corp.)

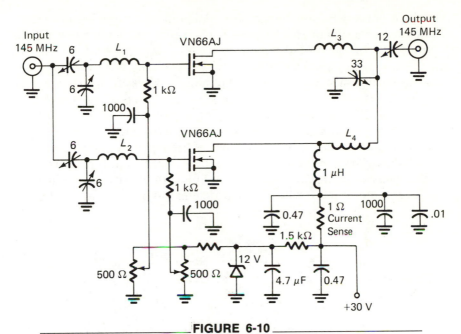

_____ **FIGURE 6-10** _____
**Two-Meter, 10-Watt PEP Linear Amplifier Using a Pair of
Power MOSFETs.**
The indirect paralleling technique essentially doubles the PEP
power available from the single device circuit of Figure 6-9.
(Courtesy of Siliconix Corp.)

in a brute-force parallel arrangement, such as would result from connecting their respective terminals together. The scheme used enables individual adjustment of optimum bias linearization of the two power MOSFETs. Additionally, better harmonic rejection is obtained than would be the case if the simpler paralleling technique were used. The inductors are the same as the corresponding ones in the single device circuit of Figure 6-9.

The 1-ohm current sense resistors in these amplifiers enable drain current to be monitored with commonly available milliammeters or voltmeters.

● **Amplitude-Modulated Amplifier: 135 MHz**

The three-stage amplifier shown in Figure 6-11 is suitable for use in the aircraft communications band where amplitude modulation has prevailed over FM and SSB modulation modes. The amplifier provides 6 W of output, and its upward modulation factor exceeds 90%. Some explanation is in order here, for the circuit is deliberately designed to enable a higher percentage of modulation to be achieved in the upward (increasing output) than the downward direction. Such an asymmetrical modulation envelope naturally produces distortion and certainly would not be suitable for the high-fidelity transmission of music. However, for voice transmission its effect on intelligibility is quite benign, and is overshadowed by the effective "talk-power" accompanying high-percentage modulation.

The salient advantage of such asymmetrical modulation is that practically full modulation capability can be realized without danger of overmodulation with its resultant *splatter* in the RF spectrum. Whereas excessive modulation in the *downward* direction is serious, upward modulation can even *exceed* 100% without such deleterious effects. Indeed, radio

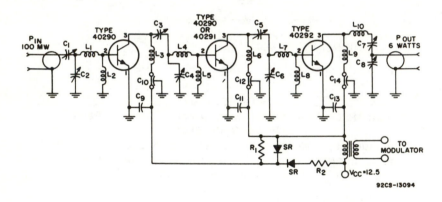

92CS-13094

$C_1, C_3, C_5,$	L_1, L_9 = 3 turns No. 16 wire,
C_7 = 3–35 pf	¼" ID, ¼" long
$C_2, C_4, C_6,$	L_2, L_5 = Ferrite choke,
C_8 = 8–60 pf	Z = 450 ohms
C_9, C_{11}, C_{13} = 0.03 μf	L_3 = RF choke, 1.5 μh
C_{10}, C_{12}, C_{14} = 1000 pf	L_4, L_7 = 4 turns No. 16 wire,
	¼" ID, ⅜" long
	L_6 = RF choke, 1.0 μh
	L_8 = wire-wound resistor,
R_1 = 220 ohms	R = 2.4 ohms
R_2 = 180 ohms	L_{10} = 5 turns No. 16 wire,
SR = 1N2858	⅜" ID, ½" long

_____ **FIGURE 6-11** _____
Amplitude Modulated Amplifier: 135 MHz.
Suitable for aircraft communications, "talk-power" is enhanced
by an asymmetrical modulation envelope wherein upward
modulation exceeds downward modulation. Minimal effect on
speech intelligibility results, and signal splatter is avoided.
(Courtesy of RCA Solid-State Division)

amateurs at one time used "exalted" modulation techniques to purposely "overmodulate" in the upward direction. This was a popular way to obtain some of the features of higher power, but at low cost. And signal splatter did not occur as long as the carrier level was not downward modulated excessively. (Too much upward modulation ultimately produces unacceptable distortion in the detectors of some receivers.)

It will be noted in Figure 6-11 that the second and third stages of the amplifier are conventionally modulated. Such tandem modulation is commonly encountered in AM solid-state amplifiers. The reason is that it is difficult to achieve high modulation factors by applying the audio modulation to the final RF power stage only. If attempted, it is generally found that the modulated transistor saturates well before 100% upward modulation can be attained. By simultaneously modulating the driver stage, the saturation region is moved up to accommodate heavier modulation.

Additionally, the first RF amplifier stage is modulated from audio derived from the diode-resistor network connected across the secondary of the modulation transformer. This network permits the first RF amplifier stage to be modulated in the upward direction but drastically limits its downward modulation. The overall result is that the RF output of the amplifying system can be safely modulated at high modulation factors without danger of overmodulation in the usual sense. (Carrier cutoff and signal splatter cannot occur.)

● Doubling the Output Power—Almost

The circuit shown in Figure 6-12 is substantially the same as that of Figure 6-11, except that the final stage consists of two parallel-connected power transistors. It will be seen that the paralleling arrangement is a bit more involved than merely connecting the respective terminals of the transistors together. Even so, it is difficult to obtain twice the output of a single transistor stage. This is primarily because of differences in the transistors. The situation is somewhat like paralleling two batteries which do not have quite the same voltage, voltage regulation, internal resistance, and temperature behavior. Another complication arises from the doubled power demand imposed upon the driver stage. Perhaps with a fortuitous choice of the RCA 40292 output transistors, one might squeeze 11-W from this arrangement.

Paralleling in higher-power output stages often involves individual networks in the *collector* as well as the base circuits. This tends to result in a closer approach to actually doubling the power available from a single transistor. (Because of the added complication, such an output stage is less frequently seen than the push-pull connection, which provides the possibility of greatly attenuating even-order harmonics.)

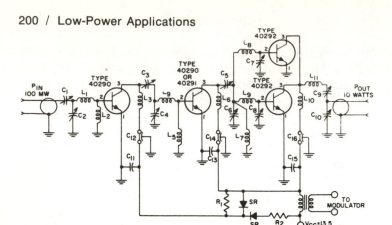

C_1, C_3, C_5, C_9 = 3–35 pf
C_2, C_4, C_6, C_{10} = 8–60 pf
C_7, C_8 = 1.5–20 pf
C_{11}, C_{13}, C_{15} = 0.03 μf
C_{12}, C_{14}, C_{16} = 1000 pf

R_1 = 33 ohms
R_2 = 36 ohms
SR = 1N2858

L_1 = 3 turns No. 16 wire,
1/4" ID, 1/4" long
L_2, L_5 = Ferrite choke,
Z = 450 ohms
L_3 = RF choke, 1.5 μh

L_4 = 4 turns No. 16 wire,
1/4" ID, 3/8" long
L_6, L_7 = RF choke, 1.0 μh
L_8, L_9 = 3 turns No. 16 wire,
1/4" ID, 3/8" long
L_{10} = 1 turn No. 16 wire,
5/16" ID, 1/8" long
L_{11} = 4 turns No. 16 wire,
3/8" ID, 1/2" long

_____ FIGURE 6-12 _____

Ten Watts AM at 135 MHz via Paralleled Output Stage.
Paralleling of RF power transistors generally does not involve the
brute-force connection of the respective terminals. In this case,
individual base-drive networks are employed.
(Courtesy of RCA Solid-State Division)

● **Complementary-Symmetry Driver Amplifier for SSB**

The broadband (2 to 30 MHz) amplifier shown in Figure 6-13 is the
equivalent of two cascaded push-pull stages even though single-ended
resonant tank circuits are used. (Such an arrangement was not feasible in
tube designs because no PNP tube was ever available.) Output power is
up to 25 W PEP, and the power gain is in the vicinity of 35 dB. As depicted, the amplifier is intended for use as a driver for a larger amplifier.
However, with the use of appropriate harmonic filters in the output, this
amplifier could also serve as a final power stage, feeding an antenna.

The dashed-line components provide an option, the use of a hybrid amplifier to increase the power gain to 45 or 50 dB. Motorola's MHW570 to MHW572 types are suitable for this application. (Gain compensation to cause roll-off below 5 MHz should be used.) The purpose of this option is to enable the amplifier to be directly driven from a balanced modulator.

The first stage, $Q1$ and $Q2$, is biased for class A operation, with quiescent collector current set at 700 mA. Q3 and Q4, the second stage, are biased for class AB operation with a quiescent collector current of about 75 mA. This biasing technique results in good even-order harmonic cancellation, as well as acceptable intermodulation distortion. Negative feedback is used with both stages. The feedback loop of the input stage contains the inductor, $L1$, and thereby produces gain compensation. This is necessary in most broadband systems to offset the tendency of transistors to develop progressively higher gain at lower frequencies.

Transformers T1 and T2 are similar, but do not have the same winding ratios. Both utilize Stackpole dual balun ferrite cores number 57-1845-24B. This core material has a permeability of 2400. The low impedance windings of both transformers consist of one turn of copper braid. The

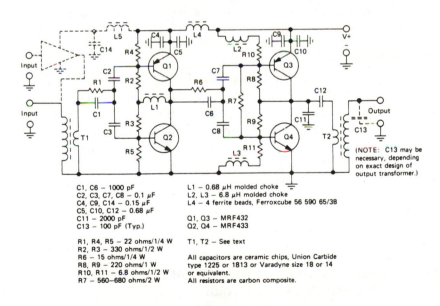

C1, C6 – 1000 pF
C2, C3, C7, C8 – 0.1 µF
C4, C9, C14 – 0.15 µF
C5, C10, C12 – 0.68 µF
C11 – 2000 pF
C13 – 100 pF (Typ.)

R1, R4, R5 – 22 ohms/1/4 W
R2, R3 – 330 ohms/1/2 W
R6 – 15 ohms/1/4 W
R8, R9 – 220 ohms/1 W
R10, R11 – 6.8 ohms/1/2 W
R7 – 560–680 ohms/2 W

L1 – 0.68 µH molded choke
L2, L3 – 6.8 µH molded choke
L4 – 4 ferrite beads, Ferroxcube 56 590 65/3B

Q1, Q3 – MRF432
Q2, Q4 – MRF433

T1, T2 – See text

All capacitors are ceramic chips, Union Carbide type 1225 or 1813 or Varadyne size 18 or 14 or equivalent.
All resistors are carbon composite.

(NOTE: C13 may be necessary, depending on exact design of output transformer.)

FIGURE 6-13

Broadband Complementary-Symmetry Amplifier for SSB.
This amplifier is capable of supplying 25-W PEP throughout the 2- to 30-MHz frequency range. It is intended for service as a driver for a larger amplifier.
(Courtesy of Motorola Semiconductor Products, Inc.)

_____ **FIGURE 6-14** _____
**Typical Ferrite Core RF Transformers Designed for Broadband
Operation.**
Close coupling is obtained by threading the wire of one winding
through the copper braid of the other. Transformers T1 and T2 of
Figure 6-13 are constructed in this manner.

primary of T1 uses two turns of AWG 22 Teflon-insulated stranded wire. And the secondary of T2 is formed of four turns of the same type of wire. In both transformers, the wire is threaded through the braid to obtain close coupling. If found more convenient, separate ferrite sleeves with the same magnetic characteristics stipulated for the dual balun structure can be used for the cores. In any event, the general construction of such transformers is illustrated in Figure 6-14.

The salient operating parameters of this amplifier are depicted by the curves in Figure 6-15. The engineering model shown in Figure 6-16 illustrates good practice in component layout.

● **Two-Watt, 2.5 GHz Oscillator Using
the Common-Collector Circuit**

Amid the controversies concerning the relative merits of common-emitter and common-base circuits for VHF, UHF, and microwave power, special transistors were developed for operation as *common-collector* oscillators. Although this configuration is not suitable for use as a stable amplifier, it is a fortuitous situation that the parasitic package capacitances

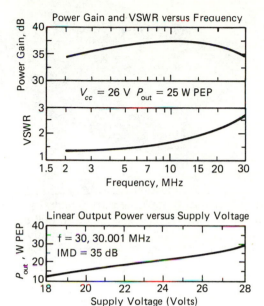

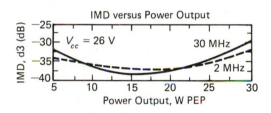

FIGURE 6-15

Operating Parameters of the Complementary-Symmetry Driver Amplifier.

Broadband performance and acceptably low intermodulation distortion merit consideration for amateur SSB applications. (Courtesy of Motorola Semiconductor Products, Inc.)

can be optimized to serve as the "voltage divider" of the Colpitts oscillator circuit. Under such conditions, it is feasible to operate the collector at ground potential for both RF and dc, thereby enhancing heat removal and circumventing RF radiation, coupling, and detuning from the heat sink.

Figure 6-17 shows the circuit of a 2-W, 2.3 GHz oscillator utilizing the TRW 62602 common-collector transistor. The approximate equivalent circuit is shown in Figure 6-18. Those familiar with tube practice will recognize the configuration as that of the Clapp series-tuned version of the Colpitts oscillator. The voltage-divider capacitances, C_{BE} and C_{CE} derive

_____ **FIGURE 6-16** _____
Engineering Model of Complementary-Symmetry Driver Amplifier.
Broadbanded for 2- to 30-MHz service, push-pull operation obtains
with simple single-ended circuitry.
(Courtesy of Motorola Semiconductor Products, Inc.)

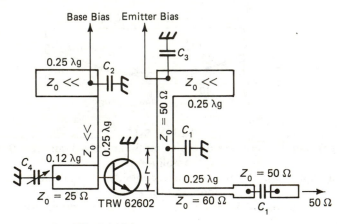

C_1 220 pF (chip)

C_2 220 pF (chip) + 10 nF

C_3 220 pF (chip) + 10 nF + 10 μF

C_4 0.6–4.5 pF (frequency tuning)

L Adjust to obtain the maximum output power

_____ **FIGURE 6-17** _____
Two-Watt, 2.3 GHz Oscillator Using Common-Collector Circuit.
The transistor is designed and packaged to deliver optimum
performance in this type of oscillator circuit.
(Courtesy of TRW)

from the transistor and its packaging. Although this concept appears simple enough, it is not always readily applicable with transistors not specifically designed for such operation. (Although appropriate values for the voltage-dividing capacitors can be attained with external chip capacitors, "ordinary" microwave transistors will not operate with their collectors at RF ground. This is because of the inductance of the collector bonding wires. Although this effect can be somewhat reduced by means of external reactance, such implementation is obviously a nuisance.)

The circuit of Figure 6-17 relies heavily on stripline transmission-line elements for simulating the function of tank inductance, RF chokes, impedance-matching transformer, and bypass capacitance. In the base circuit, the 25-ohm, one eighth-wavelength line comprises the series inductor of the frequency-determining tank; variable capacitor C_4 adjusts the frequency of the oscillator. The thin quarter-wavelength line in the base circuit serves as an RF choke, and the wide quarter-wave stud associated with it provides effective RF bypassing.

In the emitter output circuit, the line-section L cancels the emitter

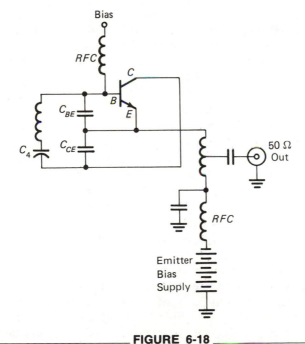

_____ **FIGURE 6-18** _____

Approximate Equivalent Circuit of the Microwave Oscillator of Figure 6-17.
The circuit is essentially a series-tuned Colpitts (Clapp) oscillator with an impedance-matching output provision. Capacitors C_{BE} and C_{CE} are parasitic packaging capacitances put to good use.

PC Board Layout for F_0 = 2.3 GHz (BW = 500 MHz)

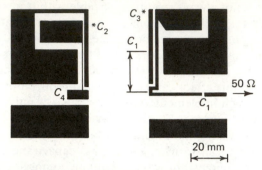

*Foil-wrap asterisked edge to ground plane.
Board material: —0.020″ glass Teflon (Er = 2.55).
Adjust L to obtain the maximum output power.

For F = 2 GHz	L = 24 mm
F = 2.3 GHz	L = 19 mm
F = 2.5 GHz	L = 14 mm

_____ FIGURE 6-19 _____

Printed Circuit Board Layout for 2.3-GHz Microwave Oscillator.
Appropriate deployment of transmission-line elements serves
functions of resonating, impedance transformation, RF choke, and
bypassing.
(Courtesy of TRW)

output capacitance. The quarter-wave, 60-ohm line is then used to transform the emitter impedance to 50 ohms. The remainder of the structure
containing L serves as RF choke and stub line for bypassing to ground.
(Because of the copper-dielectric-copper fabrication of the PC board,
there is a ground plane underneath all the stripline elements.)

Other RF circuitry features are as follows: capacitor C_1 provides dc
blocking at the output. Two C_1's are shown; the other, of identical type
and size, is a bypass capacitor. L is governed by the physical location and
connection of this capacitor.

The printed-circuit board layout is shown in Figure 6-19. The drawing
is not reproduced full scale. The widths and lengths of the stripline elements are critical. The philosophy of merely approximating geometric
shapes and dimensions is permissible for low-frequency electronic circuits, but not for microwave projects. Moreover, it is important to use
exactly the prescribed type and thickness of board material. Otherwise, it
is likely that a different dielectric constant will be encountered; this will
invalidate the dimensions. A microwave "circuit" of this type is as much
a geometrical design as it is a connection diagram.

A suitable source of base and emitter bias is shown in Figure 6-20.

This is not the thermal-tracking type used with higher-power RF circuits, but is adequate for the purpose. It provides adjustment of the operating mode between class B and class A and enables the oscillator to be self-starting. The emitter bias supply corresponds in function to the usual collector supply for common-emitter and common-base circuits. In this application, however, the supply must float with respect to the ground system of the oscillator. A voltage-regulated supply with a nominal output of 20 V and current capability of 500 mA is recommended for the emitter supply (depicted as a battery in Figure 6-20).

● Tapered-Stripline Microwave Amplifier

The simple and straightforward grounded-base amplifier shown in Figure 6-21 belies its performance capabilities: 13 W with "tank circuits" selected for 1 GHz, and 6 W when operated with 2-GHz tanks. These input and output tank circuits perform impedance matching via their tapered shapes. The relatively low impedances of both the emitter and the collector circuits are converted to 50 ohms by these tapered stripline elements. No stublines or additional reactances are needed. In principle, the transition from wide to narrow end should be by means of exponential curves. In practice, the simple trapezoidal shape often suffices.

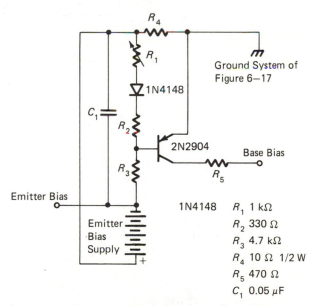

1N4148	R_1 1 kΩ
	R_2 330 Ω
	R_3 4.7 kΩ
	R_4 10 Ω 1/2 W
	R_5 470 Ω
	C_1 0.05 μF

FIGURE 6-20

Base and Emitter Bias Source for 2-Watt, 2.3-GHz Oscillator.
Note that this dc power supply must "float" with respect to the ground system of Figure 6-17.

The 2N6266 emitter-ballasted transistor operates from a 28-V dc supply and develops a collector efficiency in the vicinity of 40% at these microwave frequencies. At the same time, it is very rugged, with an infinite VSVR capability. The power levels are quite respectable for the microwave region, and good results may be anticipated for such applications as collision-avoidance systems, S-band telemetry, distance-measuring equipment, microwave-relay link, phased-array radar, and transponder.

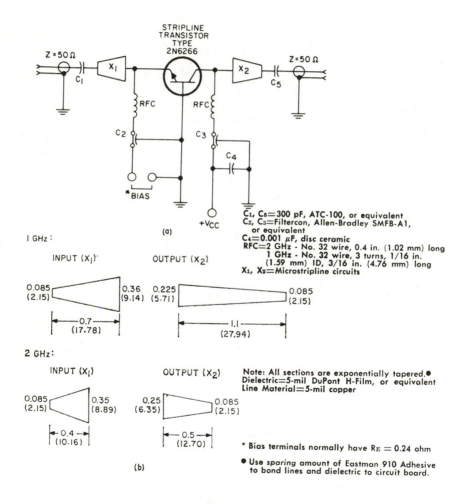

C₁, C₅=300 pF, ATC-100, or equivalent
C₂, C₃=Filtercon, Allen-Bradley SMFB-A1, or equivalent
C₄=0.001 μF, disc ceramic
RFC=2 GHz - No. 32 wire, 0.4 in. (1.02 mm) long
    1 GHz - No. 32 wire, 3 turns, 1/16 in. (1.59 mm) ID, 3/16 in. (4.76 mm) long
X₁, X₂=Microstripline circuits

Note: All sections are exponentially tapered.●
Dielectric=5-mil DuPont H-Film, or equivalent
Line Material=5-mil copper

* Bias terminals normally have R_E = 0.24 ohm

● Use sparing amount of Eastman 910 Adhesive to bond lines and dielectric to circuit board.

FIGURE 6-21

Tapered-Stripline Microwave Amplifier.

Simple impedance-matching networks are suggestive of acoustic techniques.

(Courtesy of RCA Solid-State Division)

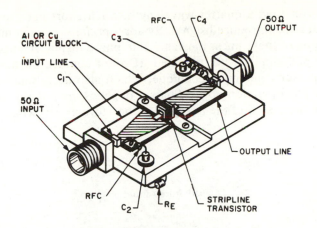

_____ **FIGURE 6-22** _____
Construction of Tapered-Stripline Microwave Amplifier.
By using appropriate stripline elements, the output power is
approximately 13 W at 1 GHz and 6 W at 2 GHz.
(Courtesy of RCA Solid-State Division)

When incorporated into amplifier chains, driver requirements can be determined from the nominal 11-dB gain developed at 1 GHz and 7-dB gain at 2 GHz.

The construction of this amplifier is shown in Figure 6-22.

WARNING: The ceramic body of this device contains beryllium oxide. Do not crush, grind, or abrade these portions because the dust resulting from such action may be hazardous if inhaled. Disposal should be by burial.

● Quadrature Amplifier Circuit

The linear amplifier shown in Figure 6-23 is essentially a doubled-up version of the circuit depicted in Figure 6-8. Of prime interest here is the combining technique which makes use of coupled transmission-line elements. The bandwidth is not as great as in the simpler single-device arrangement, but still qualifies as broadband (100 to 160 MHz). And although gain and output penalties are exacted by the 50-ohm "dummy" loads in the input and output circuits, a very considerable gain is made in the linearity of the amplifier. In this respect, such an amplifying system is superior to the parallel or push-pull combining techniques. The quadrature combining scheme utilizes transmission-line elements which are much shorter than a quarter-wavelength. Such short lines impart a 90° phase shift to an introduced signal. By appropriately cross-connecting the terminations of these lines, the same 180° phase displacements that exist in a push-pull amplifier are obtained. (Do not confuse this arrangement with a diplexer combining technique, which also uses transmission-line elements. The operation of the diplexer depends upon the quarter-wavelength characteristics of the lines and is therefore a narrowband scheme.)

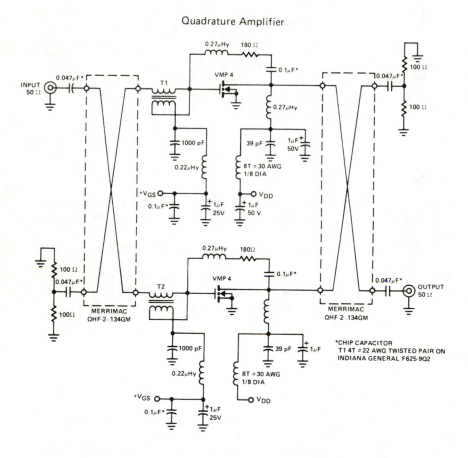

FIGURE 6-23

Quadrature Amplifier for 100- to 160-MHz Linear Service.
Power splitting at the input and power combining at the output are accomplished via coupled transmission-line elements.
(Courtesy of Siliconix Corp.)

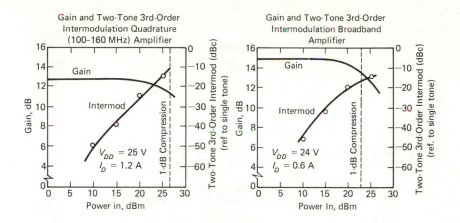

FIGURE 6-24

Gain and Intermodulation Performance of the Amplifiers Shown in Figure 6-23 and 6-8.

(a) Two power MOSFETs in quadrature circuit of Figure 6-23. (b) Single power MOSFET in broadband circuit of Figure 6-8. (Courtesy of Siliconix Corp.)

The provisions for individual gate-biasing help in adjusting for optimally low intermodulation distortion. Figures 6-24(a) and (b) compare gain and intermodulation performance of this amplifier with that of the single-device circuit of Figure 6-8. It will be seen that the quadrature amplifier provides better linearity, although it sacrifices about 3 dB of gain.

7

Medium- and High-Power Applications

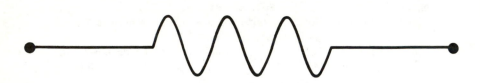

This chapter discusses practical circuits in which the output power level exceeds 25 W. Power-frequency combinations which, until recently, were "blue-sky" concepts, are shown as candidates for everyday breadboarding experiments or as well-engineered circuits for ready implementation into equipment. This should be a dynamic field during the next decade, with the major semiconductor firms vying with one another to win the competitive edge in power and frequency capability, efficiency, reliability, and cost.

● A 32-Watt Marine Band Amplifier

The 156-MHz amplifier shown in Figure 7-1 is intended for FM service. This greatly relaxes some of the stringent demands other modulation formats impose on amplifiers. The final stage comprises two 2N5996 transistors operating in parallel. The paralleling configuration is somewhat similar to one or two previously discussed. However, in this case, the two output transistors each have their individual networks. The arrangement is such that the transistors can be made to share equally in power amplification even though they do not exhibit exactly the same input and output impedances. It would be only natural to ponder why a push-pull circuit is not employed inasmuch as the circuit complexity and parts count would be about the same. The primary reason is that the salient feature of push-pull amplification, cancellation of even-order harmonics, is not always readily forthcoming in practice. (To fully exploit the potentialities of push-pull operation, it is generally desirable to use transmission-line tank circuits at VHF and to pay heed to factors affecting balanced operation. For example, divergent transistor characteristics are more difficult to cope with than in the parallel format used in this amplifier.)

It is interesting to observe that the bypassing and decoupling circuitry is as involved as the actual amplifier chain and its impedance-matching networks. This is a commonplace situation at VHF and UHF, being necessary to discourage oscillation at lower frequencies where the power gain of the stages greatly exceeds that available at the operational frequency.

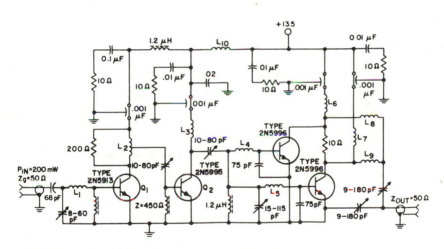

L_1 = 2 turns of No. 20 B.T., 1/4-inch diameter, 1/8 inch long
L_2 = 5 turns of No. 20 B.T., 1/4-inch diameter, 3/8 inch long, tap at 4½ turns from collector
L_3 = 5 turns of No. 20 enamel wire, 3/16-inch diameter, 1/4 inch long
L_4, L_5 = 1 turn of No. 20 B.T., 1/8-inch diameter, 1/8 inch long
L_6, L_7 = 2 turns of No. 20 B.T., 3/16-inch diameter, 1/4 inch long
L_8, L_9 = 2 turns of No. 18 B.T., 1/4-inch diameter, 3/16 inch diameter
L_{10} = 10 turns of No. 20 enamel wire, 1/4-inch diameter, close wound

FIGURE 7-1
A 32-Watt Marine Band Amplifier.
Unique paralleling technique accommodates the parameter tolerances of power transistors.
(Courtesy of RCA Solid-State Division)

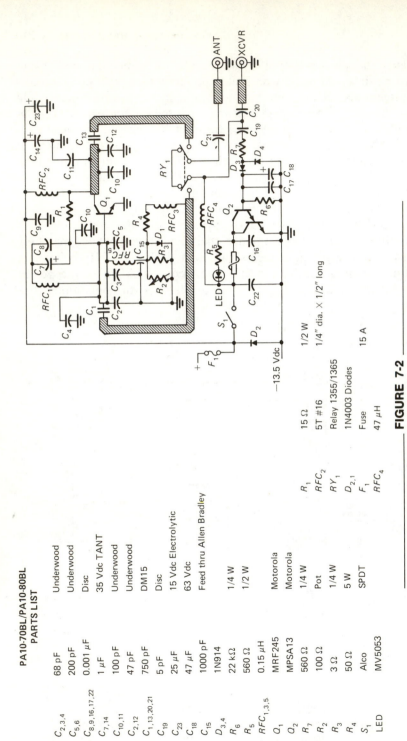

**PA10-70BL/PA10-80BL
PARTS LIST**

$C_{2,3,4}$	68 pF	Underwood
$C_{5,6}$	200 pF	Underwood
$C_{8,9,16,17,22}$	0.001 μF	Disc
$C_{7,14}$	1 μF	35 Vdc TANT
$C_{10,11}$	100 pF	Underwood
$C_{2,12}$	47 pF	Underwood
$C_{1,13,20,21}$	750 pF	DM15
C_{19}	5 pF	Disc
C_{23}	25 μF	15 Vdc Electrolytic
C_{18}	47 μF	63 Vdc
C_{15}	1000 pF	Feed thru Allen Bradley
$D_{3,4}$	1N914	
R_6	22 kΩ	1/4 W
R_5	560 Ω	1/2 W
$RFC_{1,3,5}$	0.15 μH	
Q_1	MRF245	Motorola
Q_2	MPSA13	Motorola
R_7	560 Ω	1/4 W
R_2	100 Ω	Pot
R_3	3 Ω	1/4 W
R_4	50 Ω	5 W
S_1	Alco	SPDT
LED	MV5053	
R_1	15 Ω	1/2 W
RFC_2	5T #16	1/4" dia. × 1/2" long
RY_1	Relay 1355/1365	
$D_{2,1}$	1N4003 Diodes	
F_1	Fuse	15 A
RFC_4	47 μH	

FIGURE 7-2

A 70-Watt, 2-Meter, Broadband Linear Amplifier.
With a proper antenna and RF drive source, no tuning is required.
(Courtesy of KLM Electronics, Inc.)

Such an amplifier, with integrally contained driver stages, is relatively easy to implement inasmuch as only 200 mW of input power is needed. Whereas the three cascaded stages operate "straight through" at 156 MHz, it is probable that one or more frequency multipliers will be used to process the excitation power. Such frequency-multiplying stages serve a useful purpose in FM transmitters in that they also multiply the FM deviation produced in a frequency-modulated oscillator or early buffer stage.

● A 70-Watt, 2-Meter Broadband Linear-Amplifier

The circuit shown in Figure 7-2 is that of the KLM type PA 10-70 BL linear amplifier. Microstripline techniques are used for the input and output matching networks. Additionally, the Motorola MRF245 RF power transistor has an internal low $Q\,L$ section in its base lead. The overall effect of such matching networks is that no tuning is required when the amplifier is utilized within the 143- to 149-MHz frequency range. The 70-W output rating applies for ICAS duty cycles. SSB, AM, FM, or CW operating modes can be accommodated. Input power can be in the 5- to 15-W range. The maximum input VSWR is 1.4:1. Figure 7-3 shows the underside of the amplifier. The basic simplicity of the layout cannot fail to make an impact on those whose transmitter experience derives primarily from older equipment. Here it is only necessary to connect the RF drive power, the dc operating power, and the antenna. The RF drive is intended to be provided from a transceiver.

Darlington transistor $Q2$ and its associated circuitry comprise a carrier-operated relay. With this provision, the contacts of relay RY1 activate the amplifier when the transceiver is placed in its transmit mode. This is accomplished by forward biasing the RF transistor (it operates in class AB), delivering the RF drive to its base circuit, and connecting the antenna to its collector circuit. When thus enabled, the LED in the relay circuit turns on. Conversely, when the transceiver is placed in its receive mode, the contacts of relay RY1 disconnect the forward bias from the RF transistor and connect the antenna to the transceiver. Thus, once the operator has installed the amplifier, procedures remain substantially unaffected. When the transceiver is in its receive mode, the amplifier consumes only a negligibly small direct current from the battery (on the order of 10 mA, or less). Thus, power switch S1 can be left on for long periods with no adverse effects.

Diode D1, being mounted in close thermal proximity to the RF transistor, causes the forward bias to track the operating temperature so that the 10-mA collector "idling" current remains substantially constant over the operating temperature range of the amplifier. This not only prevents

FIGURE 7-3
Underside of KLM Amplifier.
(Courtesy of KLM Electronics)

thermal runaway, but ensures that the intermodulation distortion of the amplifier remains optimally low.

Inasmuch as the basic theme of this book does not involve "how-to-construct" directions, the physical dimensions of stripline elements are not given.

● Eighty Watts of Class C Power

The uncomplicated amplifier shown in Figure 7-4 conservatively develops 80 W at 30 MHz. The operational mode is class C, and input and output impedances are 50 ohms. This amplifier is eminently suited for CW operation. The power transistor is specially designed for high-frequency service up to 30 MHz. Motorola makes two versions which differ from each other in mounting provisions. The MRF454 has a mounting flange, whereas the MRF454A has a mounting stud. Both have opposed-emitter tabs to minimize emitter lead inductance and to diminish feedback from collector to base. The packaging technique greatly contributes to stability and is one of the reasons no neutralization is needed in most implementations.

The circuit shown develops 12 dB of power gain, and collector efficiency is 50%. Of course, without a dc negative bias applied to the base, the circuit does not operate deeply into the class C region, as is commonplace with tubes. On the other hand, true class B operation would require

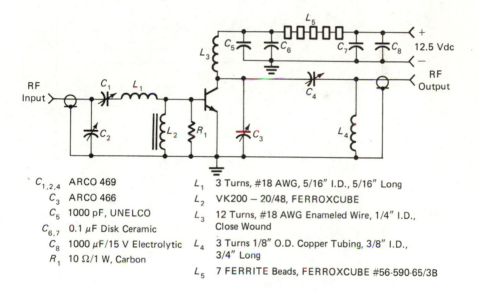

$C_{1,2,4}$	ARCO 469		L_1	3 Turns, #18 AWG, 5/16″ I.D., 5/16″ Long
C_3	ARCO 466		L_2	VK200 – 20/48, FERROXCUBE
C_5	1000 pF, UNELCO		L_3	12 Turns, #18 AWG Enameled Wire, 1/4″ I.D., Close Wound
$C_{6,7}$	0.1 μF Disk Ceramic			
C_8	1000 μF/15 V Electrolytic		L_4	3 Turns 1/8″ O.D. Copper Tubing, 3/8″ I.D., 3/4″ Long
R_1	10 Ω/1 W, Carbon			
			L_5	7 FERRITE Beads, FERROXCUBE #56-590-65/3B

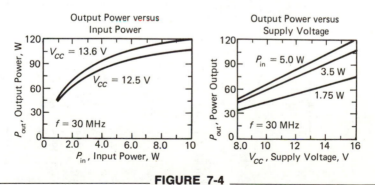

FIGURE 7-4

Conservatively Rated 80-Watt Power Amplifier for 30 MHz.
The operational mode and the intended service are suggestive of class C tube amplifiers.
(Courtesy of Motorola Semiconductor Products, Inc.)

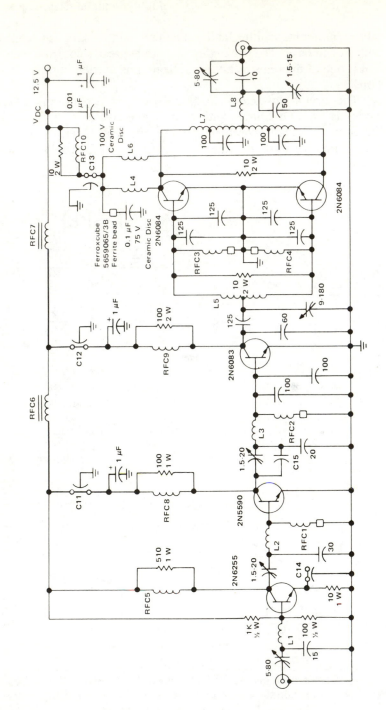

RFC1,2,3,4 — 0.15 µH, molded choke with Ferroxcube
            5659065/3B ferrite bead on ground lead
RFC5 — 0.15µH molded choke
RFC6,7 — Ferroxcube VK-200 19/4B ferrite choke
RFC8 — 4T #16 awg wire, wound on 100Ω 1 W resistor (75 nH)
RFC9 — 2T #15 awg wire, wound on 100Ω 2 W resistor (45 nH)
RFC10 — 10T #14 awg wire wound on 10Ω 2 W resistor
L1,2,3 — 1T #18 awg, ¼" dia, ¾" L (25 nH)
L4,6 — 2T #15 awg wire, ¼" dia, ½" L (30 nH)
L5,7 — See outline diagram.
L8 — #12 awg wire approximately 1" Long (9 nH)
C11,12,13 — 680 pF, Allen Bradley Type FA5C
C14 — 470 pF, Allen Bradley Type SS5D
C15 — 5 pF, Dipped Silvered Mica

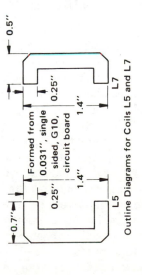

0.5"
L7
0.25"
0.25"
1.4"
1.4"
Formed from
0.031", single
sided, G10,
circuit board
0.25"
0.7"
L5
Outline Diagrams for Coils L5 and L7

FIGURE 7-5

An 80-Watt Amplifying System for Mobile 175-MHz FM Service.

Despite the resemblance of the output stage to a push-pull
configuration, it is actually a unique form of a parallel circuit.
(Courtesy of Motorola Semiconductor Products, Inc.)

forward biasing to the threshold of emitter-base conduction. Inasmuch as neither the transistor nor the circuit is expressly designed for linear amplifier service, the somewhat arbitrary class C designation is useful in suggesting that the amplifier is not intended for single-sideband service. This is not to say, however, that the experimenter cannot coax acceptable linearity from the circuit by biasing the transistor into its class AB region. But the beautiful thing about modern solid-state RF practice is that the manufacturers make optimally suited transistors available for various operational modes. Thus, there are, in addition to "workhorses," *specific* transistor families for AM, FM, SSB, CW, and driver applications. Additionally, they are grouped in frequency ranges. Such service orientation makes solid-state RF practice more of a science and less of an art.

Incidentally, it is not wise to attempt to bias an RF transistor deeply into its class C region unless the transistor has been intended for such operation. Otherwise, such reverse biasing tends to make the transistor more vulnerable to destruction from secondary breakdown or hot-spotting. That is why the base circuits of most amplifiers are designed to avoid the counterpart of self- or grid-leak biasing from the drive signal. For CW, keying is best done in a low-power driver stage.

The RF choke, L_5, comprises seven ferrite beads. A choke in this position must have the minimum inductance consistent with a reasonably high reactance at the operating frequency. This discourages the tendency of transistors to oscillate at low frequencies. For the same reason, the Q of RC choke $L2$ in the base circuit is "spoiled" by resistance $R1$.

● An 80-Watt Amplifier for Mobile 175-MHz FM Service

The amplifying system shown in Figure 7-5 incorporates unique features not commonly encountered when working with lower power levels. The first stage is biased for class A operation, whereas the subsequent stages operate in class C. This enables the system to have a very low drive requirement; only 180 mW are needed. Because of the low power level at which the first stage operates, the class A mode has minimal effect upon overall efficiency, which is nearly 50%. The approximate power levels at the input and output of the four stages are shown in Figure 7-6.

At first glance, the configuration of the output stage is suggestive of a push-pull circuit. Closer examination will reveal that the two transistors actually operate in parallel. This method of paralleling is superior to simply connecting together the respective terminals of the transistors. Although, admittedly, a little more involved, it largely circumvents or overcomes unequal load-sharing problems, as well as problems associated with inordinately low impedance. The 10-ohm resistances in the base and collector circuits dissipate only that power resulting from imbalances in these circuits. In actual practice, such wasted power is small even when

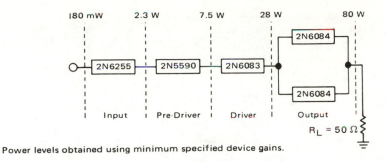

Power levels obtained using minimum specified device gains.

_____ **FIGURE 7-6** _____
Power Levels in the Four-Stage FM Amplifying System.
Note that the two output transistors operate in parallel.
(Courtesy of Motorola Semiconductor Products, Inc.)

the two transistors differ appreciably. Of course, if it is feasible to do so, it is preferable to select output transistors with closely matched characteristics. The 100-pF capacitors associated with the output tank are influential in attenuating second-harmonic energy; they are connected to the midpoints of each half of the C-shaped inductor. It is imperative that these capacitors have very low lead inductance. (Special attention to second harmonic reduction is important because, unlike a push-pull circuit, the parallel connection displays no harmonic-canceling action.)

Surprisingly clean output power is forthcoming from this amplifying system. The second harmonic is on the order of 38 dB down, and all other harmonics are greater than 50 dB down. The output stage dc current is 8.2 A when operating from a 12.5-V source. With the two 2N6084 transistors mutually mounted on a large heat sink, infinite VSWR under any phase angle will not destroy the transistors. Thus, in the extreme situations where the load may inadvertently become either short- or open-circuited, no damage will ordinarily occur.

Figure 7-7 shows an engineering model of this amplifier. The general strategy of parts layout should be noted.

● Respectable Power Levels from MOSFET Devices

The RF power amplifiers shown in Figures 7-8 and 7-9 depict the vast strides which tend to characterize electronic technology in 10-year increments. One hundred watts of VHF power from a MOSFET device was certainly a blue-sky concept a short time ago. Yet these two amplifiers are now practical realities, and the MOSFET devices are designed and packaged for such high-frequency service.

NOTE: DC power supply filtering components located
on back side of chassis —

_____ **FIGURE 7-7** _____
Engineering Model of 80-Watt, 175-MHz Amplifier.
Intended for mobile FM service, the amplifier operates from a
nominal 12.5-V dc supply.
(Courtesy of Motorola Semiconductor Products, Inc.)

Although power MOSFETs are often considered to have infinite input impedance, such a notion is valid only for dc and low-frequency applications. In the RF range, the effective input impedance is low and continues to decrease with increasing frequency. It is, however, considerably higher than for a bipolar power transistor of similar output capability. This makes the MOSFET circuit easier to translate into practical hardware, especially where the *combination* of high power and high frequency is involved. Note the difference in required input power for a given output level at 80 and 175 MHz.

The input and output impedance levels of both circuits are 50 ohms. Sufficient positive dc voltage must be applied to the V_{GG} terminal to cause a quiescent drain current of 40 mA (with no RF input). This adjusts the operating mode for class AB. Thereafter, circuit performance is straightforward, and the MOSFET is relatively immune to damage from high VSWR. (These devices are tested to withstand infinite VSWR at all phase angles at rated output power of 100 W at 175 MHz and 28-V dc supply.)

The BF100-35 RF power MOSFET used in these circuits is fabricated with beryllium oxide ceramics. This provides a low "junction"-to-case thermal resistance on the order of 0.7°C/W, and at the same time incurs negligible power dissipation. It must be borne in mind that any mechanical or chemical treatment of this material is hazardous, for the inhalation of even minute quantities of its dust or fumes can be extremely injurious

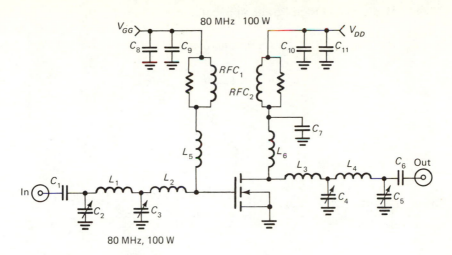

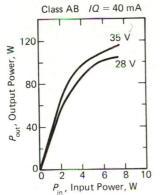

Class AB $IQ = 40$ mA

$C_{1,10}$	0.01 μF Ceramic
C_2	2–20 pF Compressed Mica
C_3	60–340 pF Compressed Mica
$C_{4,5}$	75–480 pF Compressed Mica
C_6	1000 pF Ceramic Chip
C_7	470 pF Ceramic Chip
C_8	2700 pF Mica
C_9	0.1 μF Ceramic
C_{11}	10 μF Electrolytic
L_1	2 Turns #18 Wire 3/8" I.D.
L_2	1-1/2 Turns #18 Wire 3/8" I.D.
L_3	1-3/4 Turns #18 Wire 3/8" I.D.
L_4	3-1/2 Turns #18 Wire 3/8" I.D.
L_5	8 Turns #20 Wire 1/8" I.D.
L_6	8 Turns #20 Wire 1/4" I.D.
RFC_1	6 Turns #20 Wire on Indiana General F627-8 Q_1 with 15 Ω, 1/2 W in Parallel
RFC_2	18 Turns #20 Wire on Micrometals T106-2 with 15 Ω, 2 W in Parallel

_____ **FIGURE 7-8** _____
A 100-Watt, 80-MHz Power MOSFET Amplifier.
Power MOSFET devices are now available in hitherto unthinkable
power-frequency capabilities.
(Courtesy of Communications Transistor Corporation)

175 MHz, 100 Watt

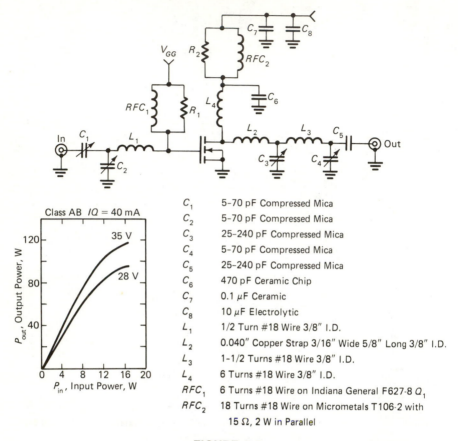

C_1	5–70 pF Compressed Mica
C_2	5–70 pF Compressed Mica
C_3	25–240 pF Compressed Mica
C_4	5–70 pF Compressed Mica
C_5	25–240 pF Compressed Mica
C_6	470 pF Ceramic Chip
C_7	0.1 μF Ceramic
C_8	10 μF Electrolytic
L_1	1/2 Turn #18 Wire 3/8" I.D.
L_2	0.040" Copper Strap 3/16" Wide 5/8" Long 3/8" I.D.
L_3	1-1/2 Turns #18 Wire 3/8" I.D.
L_4	6 Turns #18 Wire 3/8" I.D.
RFC_1	6 Turns #18 Wire on Indiana General F627-8 Q_1
RFC_2	18 Turns #18 Wire on Micrometals T106-2 with 15 Ω, 2 W in Parallel

FIGURE 7-9
A 100-Watt, 175-MHz Power MOSFET Amplifier.
Essentially a higher-frequency adaptation of the circuit of Figure 7-8; similar output power is attainable.

to health. In ordinary handling procedures where due respect is paid to the physical integrity of the device, there should be no danger.

● Implementation of the Balanced Transistor

The 125-W broadband (225 to 400 MHz) amplifier shown in Figure 7-10 utilizes a *balanced transistor*. This unique device actually comprises two RF power transistors within a common package. The overall circuit arrangement will be recognized as a push-pull configuration. Those famil-

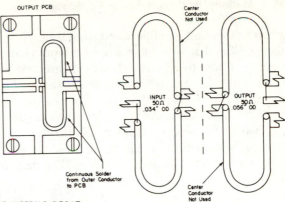

SCHEMATIC OF 125 WATT, 225-400 MHZ, WIDEBAND CIRCUIT
USING C.T.C. BAL0204-125 BALANCED TRANSISTOR

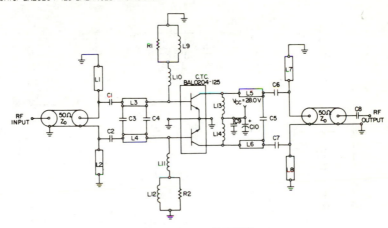

BROADBAND 225-400 MHz AMPLIFIER

BAL0204-125	CTC balanced transistor for 225 to 400 MHz operation.
C-1, C-2	39 pF ceramic chip capacitor
C-3	33 pF ceramic chip capacitor
C-4	56 pF ceramic chip capacitor
C-5	18 pF ceramic chip capacitor
C-6, C-7, C-8	27 pF ceramic chip capacitor
C-9	.1 μ F ceramic capacitor
C-10	10 μ F electrolytic capacitor
L-1, L-2, L-3, L-4, L-5, L-6, L-7, L-8	Printed on the circuit board
L-9, L-12	4.7 μ H R.F. choke
L-10, L-11, L-13, L-14	.1 μ H R.F. choke
R-1, R-2	10 Ω ¼W

FIGURE 7-10

Typical Implementation of the Balanced Transistor.

Essentially *two* RF power transistors within a common package,
this device ameliorates the two main difficulties associated with
transistor RF power circuits: inordinately low impedances and
critical emitter lead length.
(Courtesy of Communications Transistor Corp.)

225

iar with push-pull amplifiers at lower frequencies or employing vacuum tubes are familiar with the second-harmonic canceling properties of such circuitry. Often overlooked, however, are two additional features which are of prime importance in the processing of high frequencies and high powers in transistor amplifiers.

Much of the art and science involved in the practical implementation of transistor RF power circuits involves the very low input and output impedances encountered. This makes it difficult to design matching networks with practical-valued elements. The installation and wiring of such

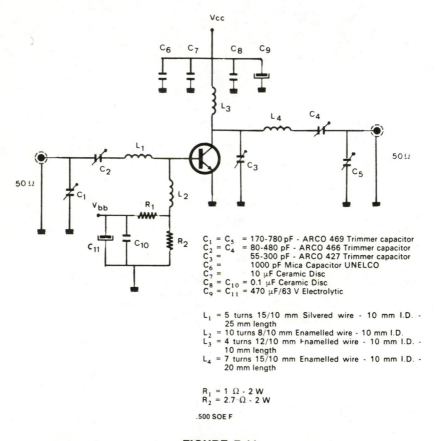

C_1 = C_5 = 170-780 pF - ARCO 469 Trimmer capacitor
C_2 = C_4 = 80-480 pF - ARCO 466 Trimmer capacitor
C_3 = 55-300 pF - ARCO 427 Trimmer capacitor
C_6 = 1000 pF Mica Capacitor UNELCO
C_7 = 10 μF Ceramic Disc
C_8 = C_{10} = 0.1 μF Ceramic Disc
C_9 = C_{11} = 470 μF/63 V Electrolytic

L_1 = 5 turns 15/10 mm Silvered wire - 10 mm I.D. - 25 mm length
L_2 = 10 turns 8/10 mm Enamelled wire - 10 mm I.D.
L_3 = 4 turns 12/10 mm Enamelled wire - 10 mm I.D. - 10 mm length
L_4 = 7 turns 15/10 mm Enamelled wire - 10 mm I.D. - 20 mm length

R_1 = 1 Ω - 2 W
R_2 = 2.7 Ω - 2 W

.500 SOE F

FIGURE 7-11
A 150-Watt, 28-MHz Test Amplifier.
Used in the manufacturer's evaluation of the high-power PT9790 transistor, this straightforward circuit has versatile application possibilities.
(Courtesy of TRW RF Semiconductors)

networks often introduces stray parameters comparable in size to the network elements themselves. A second manifestation of this basic problem is in establishing a near-zero impedance ground for the emitter (or base in common-base circuits). The penalty for a tiny fraction of an ohm of resistance and/or inductance can be severely reduced gain and various instabilities. It is these problems which the balanced transistor and its accompanying push-pull circuit action help overcome. The reason such an approach had not previously been popular is that practical considerations render it difficult to implement for the combination of high frequency and high power. The two devices must be well balanced and the RF circuitry must be compact, with vanishingly short lead lengths.

In a balanced push-pull configuration, there is *no need* to provide a good RF ground for the mutual emitter terminals (or mutual base terminals in a common-base arrangement). The circuit, itself, establishes a "virtual" RF ground at its "center of gravity." This virtual RF ground is more effective than one could derive from physical grounding. Thus, the actual grounding of the mutual emitters is for the benefit of the dc, not the RF, path. For a given power level, input and output impedances of push-pull circuits are *four* times higher than for single-ended amplifier circuits. This greatly relaxes the design limitations of the matching networks. The fact that the shunt elements of these networks do not have to be grounded also helps practical implementation. Other means than the one illustrated can be used to accomplish transformation from unbalanced to balanced circuitry, and vice versa. For example, a simple center-tapped ferrite transformer merits consideration if somewhat less emphasis is placed on broadband response.

As may be inferred, Communications Transistor Corporation also makes common-base versions of this dual device.

● A 150-Watt, 28-MHz Test Amplifier

Manufacturer's test circuits provide useful information for designers and experimenters. These circuits prove the basic performance of the device involved and are often implementable "as is," or nearly so. The high-power amplifier shown in Figure 7-11 can be operated either as a linear amplifier or in class C. The PT9790 RF transistor is electrically rugged, being designed to withstand infinite VSWR, and being rated for 300-W dissipation at 25°C. It can render good performance in SSB, FM, and AM modulation formats. The 50-V operating voltage results in an output impedance approximately 16 times that of a hypothetical like-powered device operating from a nominal 12.5-V supply. This is an important factor in the practicability of the amplifier circuit; an efficient output network for a 12.5-V device would not be easy to implement.

The input and output impedance-matching networks are similar, with L_2 and L_3 serving as RF chokes. The two chokes have different inductances to discourage self-oscillation. Inductor L_1 is silver plated to maximize its Q. Extraordinary precautions are evident in the bypassing of the power supply. The objective is to prevent oscillation or instability at low frequencies where the power gain is relatively high. The quality of the electrolytic capacitor, C_9, assumes a vital role in this regard, for one with a high ESR (effective series resistance) may not prove adequate for the purpose.

A V_{bb} source sufficient to produce a quiescent collector current of 50 mA sets this amplifier up for class AB linear operation. With no such base voltage applied, the amplifier operates class C. Special consideration is needed for the V_{bb} source. Ideally, it should track the thermal characteristics of the transistor in order to maintain the 50-mA quiescent current under actual operating conditions.

An essentially heavier duty version of the PT9790 transistor is the LOT-1000, which can develop output power on the order of 200 W.

● One Kilowatt from 2 to 30 Megahertz

Throughout the years, the power rating of 1 kW has been a target for amateur radio transmitters. Even though the technical and legal implications are different for the various modulation formats, the construction of a solid-state amplifier with 1-kW output capability is inherently a useful project where the mere novelty of solid-state power does not satisfy performance objectives.

The 1-kW amplifier to be described is convenient and economic to build, and its operation is straightforward for the following reasons:

- The amplifier is a system, consisting of four identical 300-W push-pull amplifiers.

- Output power from the four amplifiers is summed by a simple hybrid transformer arrangement. This scheme, at the same time, electrically isolates the four amplifiers from one another.

- Fifty-volt transistors are used, thereby making impedance levels practically suited for readily constructed matching transformers.

- Because of the incorporation of the driver amplifier, the 1-kW system has an overall power gain of about 34 dB, making excitation requirements as benign as with many tube "finals." Moreover, the layout and construction of the driver are very similar to that of the four 300-W amplifiers.

- Stable class AB operation prevails because of a thermally tracked bias supply.

The physical aspects of the system are shown in Figures 7-12 through 7-16. Note that pairs of 300-W amplifiers are mounted on dual heat sinks, each with its own blower.

The block diagram of the 1000-W amplifying system is shown in Figure 7-17. The format is essentially that of Figure 1-9. Although the output can approach 1200 W, there is generally some loss in the combining process and additional loss in the low-pass filter, which should be interposed between the output and the antenna. (Because of the push-pull circuitry and good linearity, the imposition made on such a filter is not great.) All things considered, the deployment of the four 300-W contributing-amplifiers enables a conservative 1000 W to be available at the antenna feeder line.

The schematic diagram of the 300-W amplifier is shown in Figure 7-18. As already pointed out, four such identical amplifiers are required in the system. Figure 7-18 is also the schematic diagram of the single driver amplifier. Although the circuitry of the 300-W amplifiers and the driver

FIGURE 7-12

Top View of 300-Watt Amplifier Board.

(Courtesy of Motorola Semiconductor Products, Inc.)

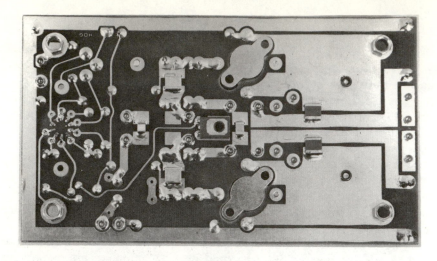

FIGURE 7-13
Bottom View of 300-Watt Amplifier Board.
(Courtesy of Motorola Semiconductor Products, Inc.)

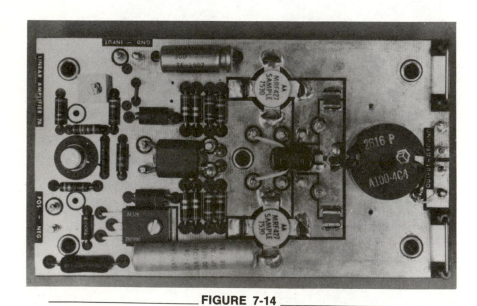

FIGURE 7-14
Driver Amplifier Board.
Key feature is similarity to 300-watt amplifier layout.
(Courtesy of Motorola Semiconductor Products, Inc.)

amplifier is the same, many of the components in the two amplifiers have different values. This, of course, is to be expected because of the disparity in power levels.

Each of the four 300-W amplifiers has its own thermal-tracking bias circuit. This is true also of the driver amplifier. The schematic of the bias source for both amplifiers is shown in Figure 7-19.

The parts list for the 300-W amplifiers and for the driver amplifier is shown in Table 7-1. The RF power transistors, $Q1$ and $Q2$, are not identified in the parts list. These transistors are the Motorola type MRF428 for the 300-W amplifiers and the Motorola type MRF427 for the 50-W driver amplifier.

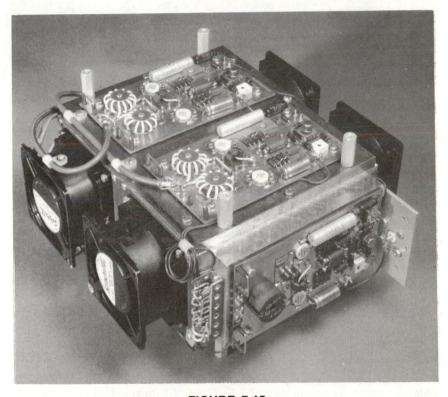

FIGURE 7-15
Pair of 300-Watt Amplifiers Such as Units A and D in Block Diagram.
Power splitter is in the foreground.
(Courtesy of Motorola Semiconductor Products, Inc.)

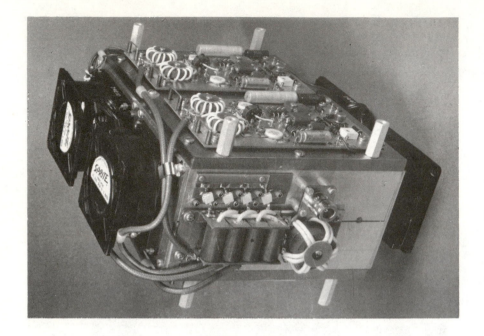

_____ **FIGURE 7-16** _____
**Pair of 300-Watt Amplifiers Such as Units B and C in Block
Diagram.**
Power combiner is in the foreground.
(Courtesy of Motorola Semiconductor Products, Inc.)

The semipictorial presentation of the output transformer in Figure
7-18 makes clear the balun nature of this transformer. It should not be construed, however, that the single-turn "windings" apply. Rather, the actual number of turns is delineated in Table 7-1. Also, the construction of
the output transformer for the driver amplifier differs from that for the
300-W amplifiers. This information, also, is given in the parts list.

It will be noted in Figure 7-18 that the secondary of input transformer
$T1$ is not center-tapped. This is permissible because base-current return
paths occur through the forward-biased base-emitter junction of the
"off" transistor. Another interesting aspect of this push-pull circuit is the
use of transformer $T2$ in place of separate RF chokes. This scheme enables the requisite reactance to be developed with relatively low I^2R loss,
an important matter at high power levels. It also allows a simple means of
providing negative feedback via tertiary winding $L5$. Decoupling and bypassing of the 50-V supply are accomplished with two separate paths involving $L3$ and $L4$, rather than by means of a mutual path, as is conventionally done. This enhances the low-frequency stability of the amplifier.

● Thermal-Tracking Bias Source

The circuit of the thermal-tracking bias source shown in Figure 7-19 is essentially a voltage regulator with a current-boosting transistor, $Q3$. A significant modification from conventional voltage regulators consists of the inclusion of "diode" $D1$ in the internal voltage-reference circuit of the MC1723 IC regulator. ($D1$ is actually the base-emitter junction of a plastic-packaged 2N5190 transistor.) $D1$ is mounted near the center of the amplifier board in close proximity to the push-pull RF power transistors, $Q1$ and $Q2$.

The voltage developed across $D1$ decreases with rising temperature, as does the *required* bias voltage of the RF power transistors. As the temperature increases, the responding junction voltage of $D1$ causes the regulator to reduce its output voltage, thereby keeping the quiescent current of the amplifier approximately constant. Such near-constancy is necessary to prevent thermal runaway and is also important in maintaining low

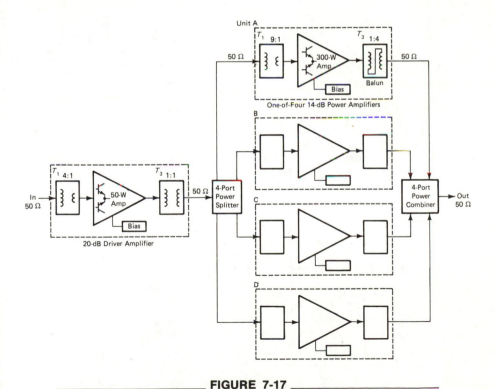

FIGURE 7-17
Block Diagram of 1-kW, 2- to 30-MHz Amplifying System.
Dashed lines indicate complete circuit boards.

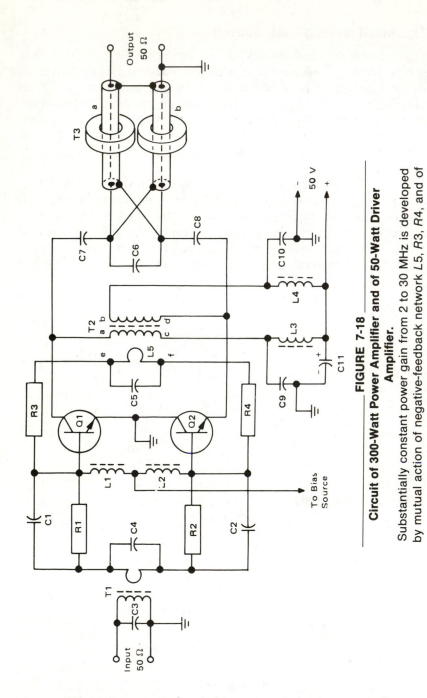

FIGURE 7-18

Circuit of 300-Watt Power Amplifier and of 50-Watt Driver Amplifier.

Substantially constant power gain from 2 to 30 MHz is developed by mutual action of negative-feedback network L5, R3, R4, and of input networks R1, C1 and R2, C2.

(Courtesy of Motorola Semiconductor Products, Inc.)

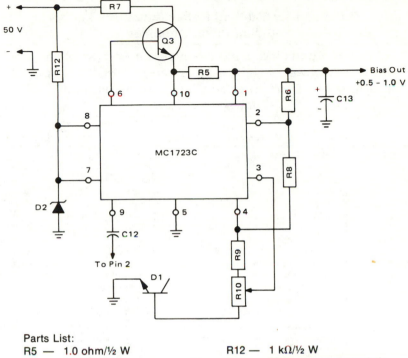

Parts List:

R5 — 1.0 ohm/½ W   R12 — 1 kΩ/½ W
R6 — 1 kΩ/½ W   C13 — 1000 μF/3 V electrolytic
R7 — 100 ohm/5 W   C12 — 1000 pF ceramic
R8 — 18 kΩ/½ W   D1 — 2N5190
R9 — 8.2 kΩ/½ W   D2 — 1N5361—1N5366
R10 — 1 kΩ trimpot   Q3 — 2N5991
R11 — ———

FIGURE 7-19

Thermal-Tracking Bias Source for 300-Watt Power Amplifiers and for Driver Amplifier.

Each amplifier board incorporates one of these bias sources. *D1*, the *sensing* diode, is the base-emitter junction of a plastic-packaged 2N5190 transistor. (Courtesy of Motorola Semiconductor Products, Inc.)

intermodulation distortion. For the latter situation, it is necessary that the bias source should have a low dynamic impedance; this is the reason the voltage-regulating circuit is used. Actually, the bias control is indirect, there being no closed loop in which the actual amplifier *current* is sensed. The reason this scheme tracks in a satisfactory manner is that the base-emitter voltages of *D1* and of the power transistors are governed by vir- ·tually the same temperature function.

_____ **TABLE 7-1** _____

Parts List for 300-Watt Power Amplifiers and for 50-Watt Driver Amplifier[a]

The circuit and parts layout are the same for both amplifier boards. The 300-W power amplifiers use a pair of MRF428 RF power transistors. The 50-W driver amplifier uses a pair of MRF427 RF power transistors.

	Power Module	Driver Amplifier
C1, C2	5600 pF	3300 pF
C3	56 pF	39 pF
C4	470 pF	Not Used
C5	560 pF	470 pF
C6	75 pF	51 pF
C7, C8	0.1 μF	0.1 μF
C9, C10	0.33 μF	0.33 μF
C11	10 μF/150 V	10 μF/150 V
R1, R2	2 x 3.9 Ω/½ W in parallel	2 x 7.5 Ω/½ W in parallel
R3, R4	2 x 6.8 Ω/½ W in parallel	2 x 18 Ω/½ W in parallel
L1, L2	Ferroxcube VK200 19/4B ferrite choke	Ferroxcube VK 200 19/4B ferrite choke
L3, L4	6 ferrite beads each, Ferroxcube 56 590 65/3B	6 ferrite beads each, Ferroxcube 56 590 65/3B
	All capacitors, except C11, are ceramic chips. Values over 100 pF are Union Carbide type 1225 or 1813 or Varadyne size 18 or 14. Others ATC Type B.	
T1	9:1 type	4:1 type
	(Ferrite cores for both: Stackpole 57-1845-24B or Fair-Rite Products 287300201 or equivalent.)	
T2	7 turns of bifilar or loosely twisted wires. (AWG #20.) Ferrite cores for both: Stackpole 57-9322, Indiana General F627-8Q1 or equivalent.	
T3	14 turns of Microdot 260-4118-00[*] 25 Ω miniature coaxial cable wound on each toroid. (Stackpole 57-9074, Indiana General F624-19Q1 or equivalent.)	11 turns of RG-196, 50 Ω miniature coaxial cable wound on a bobbin of a Ferroxcube 2616P-A100-4C4 pot core.

[*]Equivalent product is available from: W. L. Gore, Inc., Newark, Del. Part number CXN1286

[a] *(Courtesy of Motorola Semiconductor Products, Inc.)*

The nominal output voltage of the bias source is adjustable by means of potentiometer $R10$. For the 300-W amplifiers, the 25°C quiescent current is adjusted by $R10$ to be 300 mA. For the driver amplifier, the adjustment is for 80 mA. At a heat-sink temperature of 60°C, these values will hold sufficiently close so that the operating points of the RF power transistors remain substantially at their optimal class AB values. (Actually, the amplifier current decreases somewhat at the higher operating temperatures, evidencing a small amount of overcontrol. This appears to be entirely satisfactory and reinforces the conviction that there is no gradual "creep" toward thermal runaway.)

● Input Power Splitter

To avoid confusion, it should be kept in mind that all that has been hitherto discussed and described pertains to the four 300-W power amplifiers and to the 50-W driver amplifier. All subsequent discussion deals with the technique of summing up the power from the four 300-W amplifiers so that a conservative kilowatt will be available for the antenna feeder line or other 50-ohm load. The power-summing process is accomplished with the aid of a power splitter, or divider, and a power combiner. The former device splits the power from the driver amplifier into four equal parts for application to the inputs of the four 300-W amplifiers. The power combiner channels the output power from these four amplifiers into the load. Both the splitter and the combiner also *isolate* the four amplifiers from one another. A practical manifestation of this is that, should one amplifier become inoperative, the remaining three amplifiers will continue to operate at undisturbed impedance levels and the load will still receive about 750 W.

Both the power splitter and the power combiner use hybrid transformer circuits in which the transformers are transmission-line balun types. Additionally, both the splitter and combiner utilize an additional transformer for impedance matching. The power splitter will be described first.

Figure 7-20 is the schematic diagram of the power splitter. Depicted are four ferrite-bead loaded coaxial lines, an impedance-matching transformer, and four balancing resistors. The pictorial representation of the coaxial baluns is realistic; they consist of ferrite beads over a 1.2-inch length of RG-196 coaxial cable. The ferrite beads are Stackpole type 57-1511-24B, or Fair-Rite Products 2673000801, or equivalent.

The impedance-matching transformer comprises seven turns of 25-ohm miniature coaxial cable, such as Microdot 260-4118-000 on a ferrite toroid. The toroid is a Stackpole type 57-9322-11, or equivalent. This transformer is actually a transmission-line balun, itself, and the two "windings" are the inner conductor and the outer sheath of the coaxial cable. The 50-ohm impedance level from the driver amplifier is stepped down to the 12.5-ohm input impedance of the four-balun power-splitter. (Capacitor $C1$ may or may not be needed; its value can be empirically determined in terms of satisfactory input VSWR over the 2- to 30-MHz range.)

The balancing resistors (R) dissipate negligible power during normal operation of the system. In the event of failure in one or more of the 300-W power amplifiers, these resistors will dissipate varying amounts of power. If these four resistors are each 28.13 ohms, the failure of one 300-W amplifier will have little effect on overall system operation other than a reduction of power delivered to the load. The driver amplifier and the remaining three power amplifiers will otherwise operate under similar conditions to those prevailing during normal operation.

If it is deemed expedient to have such operational redundancy for the failure of *two* 300-W power amplifiers, the proper value of these balancing resistors is 25 ohms. The prospect of such a double failure might be construed a reasonable probability because of the physical arrangement of the four power amplifiers; they are constructed as two pairs of 600-W power modules.

In any event, the balancing resistors should all be equal in value and should be noninductive types. Ten-watt power ratings should suffice for most purposes. It may well be that resistance values in the neighborhood of 26.5 ohms or so could provide a compromise situation for failure of *either* one or two of the 300-W power amplifiers. Under this condition, the VSWR, linearity, and efficiency of the remaining amplifiers would be somewhat affected, but perhaps within acceptable limits.

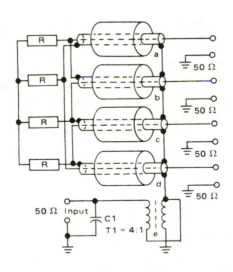

FIGURE 7-20
Four-Port Input Splitter.
The balun transmission line elements are as pictorially depicted,
lengths of coaxial cable passed through ferrite sleeves.
(Courtesy of Motorola Semiconductor Products, Inc.)

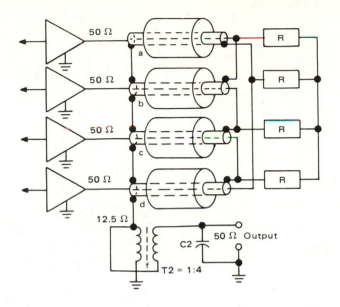

_____ **FIGURE 7-21** _____
Four-Port Output Combiner.
The balun transmission-line elements are as pictorially depicted,
lengths of coaxial cable passed through ferrite beads.
(Courtesy of Motorola Semiconductor Products, Inc.)

● **Output Power Combiner**

The output power combiner operates on similar principles to that of
the input power splitter. Power flow is in the opposite direction inasmuch
as the objective is now to channel four contributive sources of power into
the load. And because of the higher power levels involved, there are naturally differences in the construction of the elements.

Figure 7-21 is the schematic diagram of the power combiner. As with
the power splitter, we have four ferrite-sleeve loaded coaxial lines, an impedance-matching transformer, and four balancing resistors. The pictorial
representation of the coaxial baluns is again realistic; they consist of ferrite sleeves over a length of RG-142B/U coaxial cable. Each ferrite sleeve
is a Stackpole type 57-0572-27A.

Index

246